普通高等学校"十四五"规划建筑环境与能源应用工程专业精品教材

建筑环境与能源应用工程专业英语

Technical English for Built Environment and Energy Application Engineering

主编 邵璟璟
主审 [英] Jo Darkwa

华中科技大学出版社
中国·武汉

图书在版编目(CIP)数据

建筑环境与能源应用工程专业英语/邵璟璟主编. —武汉:华中科技大学出版社,2021.11
ISBN 978-7-5680-7752-1

Ⅰ.①建… Ⅱ.①邵… Ⅲ.①建筑工程-环境管理-英语-高等学校-教材 Ⅳ.①TU-023

中国版本图书馆 CIP 数据核字(2021)第 239224 号

建筑环境与能源应用工程专业英语　　　　　　　　　　　　　　邵璟璟　主编
Jianzhu Huanjing yu Nengyuan Yingyong Gongcheng Zhuanye Yingyu

策划编辑:王一洁
责任编辑:赵　萌
封面设计:原色设计
责任监印:朱　玢
出版发行:华中科技大学出版社(中国·武汉)　　电话:(027)81321913
　　　　　武汉市东湖新技术开发区华工科技园　　邮编:430223
录　　排:华中科技大学惠友文印中心
印　　刷:武汉开心印印刷有限公司
开　　本:850mm×1065mm　1/16
印　　张:10.5
字　　数:225 千字
版　　次:2021 年 11 月第 1 版第 1 次印刷
定　　价:39.80 元

本书若有印装质量问题,请向出版社营销中心调换
全国免费服务热线:400-6679-118　竭诚为您服务
版权所有　侵权必究

前　言

在"一带一路"背景下，我国建筑企业纷纷走出国门，开拓海外市场。由于我国建筑企业与国外交流频繁、交往密切，专业英语交流成为必然要求。同时，考研与出国成为当下工科毕业生的重要发展方向，而英语能力是实现个人发展不可或缺的工具，因此，提高工科学生的专业英语水平成为高等院校教学改革的重要目标。

建筑环境与能源应用工程专业英语，涉及传热学、热力学、流体力学、建筑环境学、建筑学等多个学科，专业词汇量大、逻辑复杂，如何让学生在较短的课程时间内掌握阅读理解、遣词造句的规律与方法，满足本行业工作与科研的基本要求，成为本科教学的要点与难点之一。编者结合多年的教学实践，精选阅读材料，精心组织了本书内容，便于学生在有限的时间内精读这些内容，从而快速提高专业英语综合素质。

本书包括5个部分12个单元，内容涉及建筑环境与能源应用工程专业基础、暖通空调工程、建筑给排水、建筑电气与自动控制等方面的英语阅读理解以及专业写作等。本书从专业用语、系统设计、文本结构、设计汇报四方面进行内容设计，有助于提高读者的听说读写能力。本书中的专业英语素材均来自国外的大学及专业设计院。每单元分为课前热身、专业阅读和课后扩展三部分，便于教师组织课堂教学内容。每单元配有相关知识点视频，视频内容由编写团队录制，内容多样，有设备选型、系统设计和设计规范阅读等，便于学生学习。

本书由宁波工程学院邵璟璟主编，诺丁汉大学（University of Nottingham）Jo Darkwa教授主审，书中视频由宁波工程学院邵璟璟、郭秀娟、魏莉莉、吴宏伟录制。

由于时间仓促，加之教材涉及的内容较广泛，书中难免出现错漏，希望广大读者不吝指正，以便今后不断改进和完善。

编者
2021年7月

教学支持说明

普通高等学校"十四五"规划建筑环境与能源应用工程专业精品教材系华中科技大学出版社重点规划的系列教材。

为了提高教材的使用效率,满足高校授课教师的教学需求,更好地提供教学支持,本教材配备了相应的教学资料(PPT电子教案、教学大纲等)和拓展资源(案例库、习题库、试卷库、视频资料等)。

我们将向使用本教材的高校授课教师免费赠送相关教学资源,烦请授课教师通过电话、QQ、邮件或加入土建专家俱乐部QQ群等方式与我们联系。

联系方式:

地址:湖北省武汉市东湖新技术开发区华工园六路华中科技大学出版社

邮编:430223

电话:027-81339688 转 782

QQ:61666345

E-mail:wangyijie027@163.com

土建专家俱乐部 QQ 群:947070327

土建专家俱乐部 QQ 群二维码:

土建专家俱乐部 QQ 群作为资源共享、专业交流、经验分享的平台,欢迎您的加入!

Contents

Part I Fundamental Sciences

Unit 1 Fundamentals of Thermal Sciences ·· (2)
    Warm-up Activities ·· (2)
    Intensive Reading ·· (3)
    Extensive Reading ··· (7)

Unit 2 Fluid Mechanics ·· (16)
    Warm-up Activities ·· (16)
    Intensive Reading ·· (17)
    Extensive Reading ··· (20)

Unit 3 Built Environment ··· (28)
    Warm-up Activities ·· (28)
    Intensive Reading ·· (29)
    Extensive Reading ··· (33)

Part II Heating Ventilation and Air Conditioning

Unit 4 Heating ·· (42)
    Warm-up Activities ·· (42)
    Intensive Reading ·· (43)
    Extensive Reading ··· (47)

Unit 5 Ventilation ·· (54)
    Warm-up Activities ·· (54)
    Intensive Reading ·· (54)
    Extensive Reading ··· (57)

Unit 6 Air Conditioning ··· (65)
    Warm-up Activities ·· (65)
    Intensive Reading ·· (66)
    Extensive Reading ··· (71)

Part III Building Water Supply and Drainage

Unit 7 Building Water Supply ··· (82)

　　　　Warm-up Activities ⋯⋯⋯⋯⋯⋯⋯⋯⋯⋯⋯⋯⋯⋯⋯⋯⋯⋯⋯⋯⋯⋯⋯⋯⋯⋯⋯⋯ (82)
　　　　Intensive Reading ⋯⋯⋯⋯⋯⋯⋯⋯⋯⋯⋯⋯⋯⋯⋯⋯⋯⋯⋯⋯⋯⋯⋯⋯⋯⋯⋯⋯⋯ (84)
　　　　Extensive Reading ⋯⋯⋯⋯⋯⋯⋯⋯⋯⋯⋯⋯⋯⋯⋯⋯⋯⋯⋯⋯⋯⋯⋯⋯⋯⋯⋯⋯⋯ (88)
　　Unit 8　Building Drainage System ⋯⋯⋯⋯⋯⋯⋯⋯⋯⋯⋯⋯⋯⋯⋯⋯⋯⋯⋯⋯⋯⋯⋯ (94)
　　　　Warm-up Activities ⋯⋯⋯⋯⋯⋯⋯⋯⋯⋯⋯⋯⋯⋯⋯⋯⋯⋯⋯⋯⋯⋯⋯⋯⋯⋯⋯⋯ (94)
　　　　Intensive Reading ⋯⋯⋯⋯⋯⋯⋯⋯⋯⋯⋯⋯⋯⋯⋯⋯⋯⋯⋯⋯⋯⋯⋯⋯⋯⋯⋯⋯⋯ (95)
　　　　Extensive Reading ⋯⋯⋯⋯⋯⋯⋯⋯⋯⋯⋯⋯⋯⋯⋯⋯⋯⋯⋯⋯⋯⋯⋯⋯⋯⋯⋯⋯⋯ (99)

Part IV　Electricity

Unit 9　Electrical Power Systems in Buildings ⋯⋯⋯⋯⋯⋯⋯⋯⋯⋯⋯⋯⋯⋯⋯ (108)
　　Warm-up Activities ⋯⋯⋯⋯⋯⋯⋯⋯⋯⋯⋯⋯⋯⋯⋯⋯⋯⋯⋯⋯⋯⋯⋯⋯⋯⋯⋯⋯⋯⋯ (108)
　　Intensive Reading ⋯⋯⋯⋯⋯⋯⋯⋯⋯⋯⋯⋯⋯⋯⋯⋯⋯⋯⋯⋯⋯⋯⋯⋯⋯⋯⋯⋯⋯⋯⋯ (109)
　　Extensive Reading ⋯⋯⋯⋯⋯⋯⋯⋯⋯⋯⋯⋯⋯⋯⋯⋯⋯⋯⋯⋯⋯⋯⋯⋯⋯⋯⋯⋯⋯⋯ (113)
Unit 10　Building Automation System ⋯⋯⋯⋯⋯⋯⋯⋯⋯⋯⋯⋯⋯⋯⋯⋯⋯⋯⋯⋯⋯ (121)
　　Warm-up Activities ⋯⋯⋯⋯⋯⋯⋯⋯⋯⋯⋯⋯⋯⋯⋯⋯⋯⋯⋯⋯⋯⋯⋯⋯⋯⋯⋯⋯⋯⋯ (121)
　　Intensive Reading ⋯⋯⋯⋯⋯⋯⋯⋯⋯⋯⋯⋯⋯⋯⋯⋯⋯⋯⋯⋯⋯⋯⋯⋯⋯⋯⋯⋯⋯⋯⋯ (122)
　　Extensive Reading ⋯⋯⋯⋯⋯⋯⋯⋯⋯⋯⋯⋯⋯⋯⋯⋯⋯⋯⋯⋯⋯⋯⋯⋯⋯⋯⋯⋯⋯⋯ (126)

Part V　Writing Like a Professional

Unit 11　Technical Writing ⋯⋯⋯⋯⋯⋯⋯⋯⋯⋯⋯⋯⋯⋯⋯⋯⋯⋯⋯⋯⋯⋯⋯⋯⋯⋯⋯ (134)
Unit 12　Design Specification ⋯⋯⋯⋯⋯⋯⋯⋯⋯⋯⋯⋯⋯⋯⋯⋯⋯⋯⋯⋯⋯⋯⋯⋯⋯ (140)

Appendixes　Important Words and Phrases ⋯⋯⋯⋯⋯⋯⋯⋯⋯⋯⋯⋯⋯⋯⋯⋯⋯ (144)
　　基础科学中英文对照词汇表 ⋯⋯⋯⋯⋯⋯⋯⋯⋯⋯⋯⋯⋯⋯⋯⋯⋯⋯⋯⋯⋯⋯⋯⋯⋯ (144)
　　HVACR 中英文对照词汇表 ⋯⋯⋯⋯⋯⋯⋯⋯⋯⋯⋯⋯⋯⋯⋯⋯⋯⋯⋯⋯⋯⋯⋯⋯⋯⋯ (148)
　　给排水中英文对照词汇表 ⋯⋯⋯⋯⋯⋯⋯⋯⋯⋯⋯⋯⋯⋯⋯⋯⋯⋯⋯⋯⋯⋯⋯⋯⋯⋯ (154)
　　电气工程中英文对照词汇表 ⋯⋯⋯⋯⋯⋯⋯⋯⋯⋯⋯⋯⋯⋯⋯⋯⋯⋯⋯⋯⋯⋯⋯⋯⋯ (156)
References ⋯⋯⋯⋯⋯⋯⋯⋯⋯⋯⋯⋯⋯⋯⋯⋯⋯⋯⋯⋯⋯⋯⋯⋯⋯⋯⋯⋯⋯⋯⋯⋯⋯⋯⋯⋯ (158)

Part I
Fundamental Sciences

Unit 1 Fundamentals of Thermal Sciences

Warm-up Activities

1. Learn the following words and find relevant information.

                                Chinese   **Definition**

(1) Carnot cycle
(2) dew-point
(3) enthalpy
(4) equilibrium
(5) heat
(6) irreversible cycle
(7) psychrometric chart
(8) Rankine cycle
(9) saturated air
(10) specific heat
(11) state
(12) superheated steam
(13) system
(14) temperature
(15) work
(16) boundary layer
(17) heat transfer rate
(18) Reynolds number
(19) Nusselt number
(20) thermal resistance

2. Oral Exercise: What is the importance of heat?

Intensive Reading

What Is Thermal Science

The word thermal stems from the Greek word *therme*, which means heat. Therefore, thermal sciences can loosely be defined as the sciences that deal with heat [1]. The recognition of different forms of energy and its transformations has forced this definition to be broadened. Today, the physical sciences that deal with energy and the transfer, transport, and conversion of energy *are usually referred to as* thermal sciences. Traditionally, the thermal sciences are studied under the subcategories of thermodynamics and heat transfer [2].

The design and analysis of most thermal systems such as power plants, automotive engines, and refrigerators involve all categories of thermal sciences as well as other sciences. For example, designing the radiator of a car involves the determination of the amount of energy transfer from a knowledge of the properties of the coolant using thermodynamics, and the determination of the size and shape of the inner tubes and the outer fins using heat transfer. Once the basic principles are mastered, they can then be synthesized by solving comprehensive real-world practical problems [3].

Application Areas of Thermal Sciences

All activities in nature involve some interaction between energy and matter; thus it is hard to imagine an area that does not relate to thermal sciences in some manner. Therefore, developing a good understanding of basic principles of thermal sciences has long been an essential part of engineering education. Thermal sciences are commonly encountered in many engineering systems and other aspects of life. An ordinary house is, in some respects, an exhibition hall filled with wonders of thermal sciences. Many ordinary household utensils and appliances are designed, in whole or in part, by using the principles of thermal sciences. Some examples include the electric or gas range, heating and air-conditioning systems, refrigerator, humidifier, pressure cooker, water heater, shower, iron, plumbing and sprinkling systems, and even the computer, TV, and DVD player.

On a larger scale, thermal sciences play a major part in the design and analysis of automotive engines, rockets, jet engines, and conventional or nuclear power plants, solar collectors, the transportation of water, crude oil, and natural gas, the water distribution systems in cities, and the design of vehicles from ordinary cars to airplanes. The energy-efficient home that you may be living in, for example, is designed on the basis of minimizing heat loss in winter and heat gain in summer. The

---

thermal sciences
热物理

subcategory
分支

coolant
冷却液

energy and matter
能量和物质
basic principle
基本原则
heating
供暖
air-conditioning
空调
refrigerator
制冷机
humidifier
加湿器
plumbing
管道
sprinkling systems
喷淋系统

size, location, and the power input of the fan of your computer are also selected after a thermodynamic, heat transfer, and fluid flow analysis of the computer.

Thermodynamics

Thermodynamics can be defined as the science of energy. Although everybody has a feeling of what energy is, it is difficult to give a precise definition for it. Energy can be viewed as the ability to cause changes [4]. The name thermodynamics stems from the Greek words therme (heat) and dynamic (power), which is most descriptive of the early efforts to convert heat into power. Today the same name is broadly interpreted to include all aspects of energy and energy transformations including power generation, refrigeration, and relationships among the properties of matter.

One of the most fundamental laws of nature is the conservation of energy principle. It simply states that during an interaction, energy can change from one form to another but the total amount of energy remains constant. That is, energy cannot be created or destroyed. A rock falling off a cliff, for example, picks up speed as a result of its potential energy being converted to kinetic energy. The conservation of energy principle also forms the backbone of the diet industry: A person who has a greater energy input (food) than energy output (exercise) will gain weight (store energy in the form of fat), and a person who has a smaller energy input than output will lose weight. The change in the energy content of a body or any other system is equal to the difference between the energy input and the energy output, and the energy balance is expressed as $E_{in} - E_{out} = \Delta E$.

The First Law of Thermodynamics is simply an expression of the conservation of energy principle, and it asserts that energy is a thermodynamic property. The Second Law of Thermodynamics asserts that energy has quality as well as quantity, and actual processes occur in the direction of decreasing quality of energy [5]. For example, a cup of hot coffee left on a table eventually cools, but a cup of cool coffee in the same room never gets hot by itself. The high-temperature energy of the coffee is degraded (transformed into a less useful form at a lower temperature) once it is transferred to the surrounding air.

Although the principles of thermodynamics have been in existence since the creation of the universe, thermodynamics did not emerge as a science until the construction of the first successful atmospheric steam engines in England by Thomas Savery in 1697 and Thomas Newcomen in 1712. These engines were very slow and inefficient, but they opened the way for the development of a new science. The First and Second Laws of Thermodynamics emerged simultaneously in the 1850s, primarily out of the works of William Rankine, Rudolph Clausius, and Lord Kelvin.

The term thermodynamics was first used in a publication by Lord Kelvin in 1849. The first thermodynamics textbook was written in 1859 by William Rankine, a professor at the University of Glasgow. It is well-known that a substance consists of a large number of particles called molecules.

The properties of the substance naturally depend on the behaviour of these particles. For example, the pressure of a gas in a container is the result of momentum transfer between the molecules and the walls of the container. However, one does not need to know the behaviour of the gas particles to determine the pressure in the container. It would be sufficient to attach a pressure gage to the container. This macroscopic approach to the study of thermodynamics that does not require a knowledge of the behaviour of individual particles is called classical thermodynamics [6]. It provides a direct and easy way to the solution of engineering problems. A more elaborate approach, based on the average behaviour of large groups of individual particles, is called statistical thermodynamics.

Heat Transfer

A cold canned drink left in a room warms up and a warm canned drink left in a refrigerator cools down. This is accomplished by the transfer of energy from the warm medium to the cold one. The energy transfer is always from the higher temperature medium to the lower temperature one, and the energy transfer stops when the two mediums reach the same temperature. Energy exists in various forms. Heat transfer primarily interests in heat, which is the form of energy that can be transferred from one system to another as a result of temperature difference [7]. The science that deals with the determination of the rates of such energy transfers is heat transfer.

You may be wondering why a detailed study is undertaken on heat transfer. After all, the amount of heat transfer for any system undergoing any process can be determined using a thermodynamic analysis alone. The reason is that thermodynamics is concerned with the amount of heat transfer as a system undergoes a process from one equilibrium state to another, and it gives no indication about how long the process will take. A thermodynamic analysis simply tells how much heat must be transferred to realize a specified change of state to satisfy the conservation of energy principle. In practice, the rate of heat transfer (heat transfer per unit time) is more an issue than the amount of it. For example, determine the amount of heat transferred from a thermos bottle as the hot coffee inside cools from 90℃ to 80℃ by a thermodynamic analysis alone. But a typical user or designer of a thermos bottle is primarily interested in how long it will be before the hot coffee inside cools to 80℃, and a thermodynamic analysis cannot answer this question.

momentum
动量

macroscopic
宏观的

statistical thermodynamics
统计热力学

equilibrium
平衡

Determining the rates of heat transfer to or from a system and thus the times of heating or cooling, as well as the variation of the temperature, is the subject of heat transfer. Thermodynamics deals with equilibrium states and changes from one equilibrium state to another. Heat transfer, on the other hand, deals with systems that lack thermal equilibrium, and thus it is a non-equilibrium phenomenon. Therefore, the study of heat transfer cannot be based on the principles of thermodynamics alone.

However, the laws of thermodynamics lay the framework for the science of heat transfer. The First Law requires that the rate of energy transfer into a system be equal to the rate of increase of the energy of that system. The Second Law requires that heat be transferred in the direction of decreasing temperature. This is like a car parked on an inclined road that must go downhill in the direction of decreasing elevation when its brakes are released. It is also analogous to the electric current flowing in the direction of decreasing voltage or the fluid flowing in the direction of decreasing total pressure. The basic requirement for heat transfer is the presence of a temperature difference. There can be no net heat transfer between two bodies that are at the same temperature. The temperature difference is the driving force for heat transfer, just as the voltage difference is the driving force for electric current flow and pressure difference is the driving force for fluid flow. The rate of heat transfer in a certain direction depends on the magnitude of the temperature gradient (the temperature difference per unit length or the rate of change of temperature) in that direction. The larger the temperature gradient, the higher the rate of heat transfer.

analogous
相似的

temperature gradient
温度梯度

Notes

[1] Thermal一词源于希腊语 therme,意思是热。因此,不是很严谨地说,热科学是一门研究热的科学。

[2]传统的热科学是从两个子范畴进行研究的:热力学和传热学。

[3]一旦掌握了基本原理,就可以融会贯通解决现实世界的实际问题。

[4]热力学可以被定义为一门研究能量的科学。虽然每个人都能感知到能量,但很难给能量下一个精确的定义。能量可以被看作引起变化的能力。

[5]热力学第一定律是能量守恒原理的一个简单的表达,它断言能量是热力学性质。热力学第二定律认为能量既有质量又有数量,而且实际过程是朝着降低能量的方向发生的。

[6]这种不需要了解单个粒子的行为的研究热力学的宏观方法,称为经典热力学。

[7]传热主要关注的是热。热是一种能量形式。由于温差的存在,热可以从一个系统传递到另一个系统。

Extensive Reading

Heat Transfer

Heat Transfer

The science of thermodynamics deals with the amount of heat transfer, as a system undergoes a process from one equilibrium state to another, and makes no reference to how long the process will take. But in engineering, we are often interested in the rate of heat transfer, which is the topic of the science of heat transfer. Heat transfer studies energy in transit including the relationship between energy, matter, space and time. The three principal modes of heat transfer examined are conduction, convection and radiation, where all three modes are affected by the thermophysical properties, geometrical constraints and the temperatures associated with the heat sources and sinks used to drive heat transfer. In general, heat transfer describes the flow of heat (thermal energy) due to temperature differences and the subsequent temperature distribution and changes.

Conduction

Conduction is the transfer of energy from the more energetic particles of a substance to the adjacent, less energetic ones as a result of interactions between the particles. It is the simplest heat transfer model in terms of being able to create a mathematical explanation for what's happening. It is the movement of kinetic energy in materials from higher temperature areas to lower temperature areas through a substance. The molecules will simply give their energy to adjacent molecules until an equilibrium is reached. Conduction models do not deal with the movement of particles within the material.

Consider steady heat conduction through a large plane wall of thickness $\Delta x = L$ and area A, as shown in Fig. 1-1. The temperature difference across the wall is $\Delta T = T_2 - T_1$. Experiments have shown that the rate of heat transfer Q through the wall is doubled when the temperature difference ΔT across the wall or the area A normal to the direction of heat transfer is doubled, but is halved when the wall thickness L is doubled.

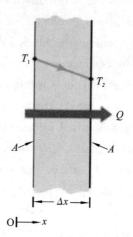

Figure 1-1 Heat conduction through a large plane wall of thickness Δx and area A

Thus the rate of heat conduction through a plane layer is proportional to the temperature difference across the layer and the heat transfer area, but is inversely proportional to the thickness of the layer. The constant of proportionality k is the thermal conductivity of the material, which is a measure of the ability of a material to conduct heat. The total heat conducted can be calculated with Fourier's Law of Heat Conduction after J. Fourier, who expressed it first in his heat transfer text in 1822.

$$Q = kA \frac{T_1 - T_2}{L} \qquad (1)$$

Convection

Convection is the mode of heat transfer between a solid surface and the adjacent liquid or gas that is in motion, and it involves the combined effects of conduction and fluid motion. It concerns the heat transfer through fluid (like air or water) motion. The difference between conduction and convection is the motion of a material carrier; convection is the movement of the thermal energy by moving hot fluid. Usually this motion occurs as a result of differences in density. Warmer particles are less dense, so particles with higher temperature will move to regions where the temperature is cooler and the particles with lower temperature will move to areas of higher temperature. The fluid will remain in motion until equilibrium is reached.

The faster the fluid motion is, the greater the convection heat transfer. In the absence of any bulk fluid motion, heat transfer between a solid surface and the adjacent fluid is by pure conduction. The presence of bulk motion of the fluid enhances the heat transfer between the solid surface and the fluid, but it also complicates the determination of heat transfer rates. Consider the cooling of a hot block by blowing cool air over its top surface. Heat is first transferred to the air layer adjacent to the block by conduction. This heat is then carried away from the surface by convection, that is, by the combined effects of conduction within the air that is due to random motion of air molecules and the bulk or macroscopic motion of the air that removes the heated air near the surface and replaces it by the cooler air. Convection is called forced convection, if the fluid is forced to flow over the surface by external means such as a fan, pump, or the wind. In contrast, convection is called natural (or free) convection, if the fluid motion is caused by buoyancy forces that are induced by density differences due to the variation of temperature in the fluid.

Despite the complexity of convection, the rate of convection heat transfer is observed to be proportional to the temperature difference, and is conveniently expressed by Newton's Law of Cooling as

$$Q = hA(T_2 - T_{f2}) \qquad (2)$$

where h is the convection heat transfer coefficient in $W/(m^2 \cdot K)$. A is the surface area through which convection heat transfer takes place, T_2 is the surface temperature, and T_{f2} is the temperature of the fluid sufficiently far from the surface. The convection heat transfer coefficient h is not a property of the fluid. It is an experimentally determined parameter whose value depends on all the variables influencing convection such as the surface geometry, the nature of fluid motion, the properties of the fluid, and the bulk fluid velocity.

Radiation

Radiation is the energy emitted by matter in the form of electromagnetic waves (or photons) as a result of the changes in the electronic configurations of the atoms or molecules.

In heat transfer studies we are interested in thermal radiation, which is the form of radiation emitted by bodies because of their temperature. It differs from other forms of electromagnetic radiation such as X-rays, gamma rays, microwaves, radio waves, and television waves that are not related to temperature. All bodies at a temperature above absolute zero emit thermal radiation. Radiation is a volumetric phenomenon, and all solids, liquids, and gases emit, absorb, or transmit radiation to varying degrees. However, radiation is usually considered to be a surface phenomenon for solids that are opaque to thermal radiation such as metals, wood, and rocks since the radiation emitted by the interior regions of such material can never reach the surface, and the radiation incident on such bodies is usually absorbed within a few micrometres from the surface. The maximum rate of radiation that can be emitted from a surface at a thermodynamic temperature T (in K) is given by the Stefan-Boltzmann Law as

$$Q = A \varepsilon \sigma T^4 \tag{3}$$

where $\sigma = 5.670 \times 10^{-8}\ W/(m^2 \cdot K^4)$, is the Stefan-Boltzmann constant. ε is the emissivity of the surface. The property emissivity, whose value is in the range $0 \leqslant \varepsilon \leqslant 1$, is a measure of how closely a surface approximates a blackbody for which $\varepsilon = 1$.

⟫⟫⟫ Post-Reading Exercises

1. Fill in the blanks.

1: _____
2: _____
3: _____
4: _____

2. Translate the following sentences into Chinese.

(1) If a process occurs in an isolated system, the entropy of the system increases for irreversible processes and remains constant for reversible processes. It never decreases.

(2) The heat transfer characteristics of a solid material are measured by a property called the thermal conductivity (k). It is a measure of a substance's ability to transfer heat through a solid by conduction. The thermal conductivity of most liquids and solids varies with temperature. For vapours, it depends upon pressure.

(3) The convection heat transfer coefficient (h), defines the heat transfer due to convection. The convection heat transfer coefficient is sometimes referred to as a film coefficient and represents the thermal resistance of a relatively stagnant layer of fluid between a heat transfer surface and the fluid medium.

3. Please describe the following diagram with no less than 100 words. (Tips: You can start with rephrasing the paragraph in text. Do not change the meaning, but with your own words. It is the first step in academic writing.)

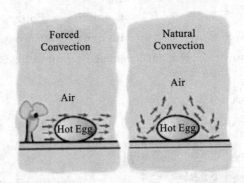

Thermodynamics

Thermodynamics studies energy, energy transformations and its relation to matter. The analysis of thermal systems is achieved through the application of the governing conservation equations, namely Conservation of Mass, Conservation of Energy (the First Law of Thermodynamics), the Second Law of Thermodynamics and the property relations. It is macroscopic variables such as internal energy, entropy, and pressure, that describe a body of matter or radiation. It states that the behaviour of those variables is subject to general constraints, which are common to all materials, not the peculiar properties of particular materials. These general constraints are expressed in the three laws of thermodynamics.

The Zeroth Law of Thermodynamics

Statement: If two thermodynamic systems are each in thermal equilibrium with a third one, then they are in thermal equilibrium with each other. Accordingly, thermal equilibrium between systems is a transitive relation.

Although everyone is familiar with temperature as a measure of "hotness" or "coldness", it is not easy to give an exact definition for it. Based on physiological sensations, people express the level of temperature qualitatively with words like freezing cold, cold, warm, hot, and red-hot.

However, numerical values cannot be assigned to temperatures based on sensations alone. Furthermore, senses may be misleading. A metal chair, for example, will feel much colder than a wooden one even when both are at the same temperature. Fortunately, several properties of materials change with temperature in a repeatable and predictable way, and this forms the basis for accurate temperature measurement. The commonly used mercury-in-glass thermometer, for example, is based on the expansion of mercury with temperature. Temperature is also measured by using several other temperature-dependent properties. It is a common experience that a cup of hot coffee left on the table eventually cools off and a cold drink eventually warms up. That is, when a body is brought into contact with another body that is at a different temperature, heat is transferred from the body at higher temperature to the one at lower temperature until both bodies attain the same temperature (Fig. 1-2). At that point, the heat transfer stops, and the two bodies are said to have reached thermal equilibrium.

The equality of temperature is the only requirement for thermal equilibrium. The Zeroth Law of Thermodynamics may seem silly that such an obvious fact is called one of the basic laws of thermodynamics. However, it cannot be concluded from the other laws of thermodynamics, and it serves as a basis for the validity of

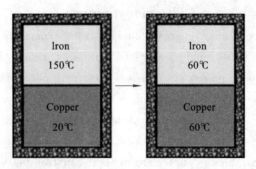

Figure 1-2 Two bodies reaching thermal equilibrium after being brought into contact in an isolated enclosure

temperature measurement. By replacing the third body with a thermometer, the Zeroth Law can be restated as two bodies are in thermal equilibrium if both have the same temperature reading even if they are not in contact. The Zeroth Law was first formulated and labelled by Fowler in 1931. As the name suggests, its value as a fundamental physical principle was recognized more than half a century after the formulation of the First and the Second Laws of Thermodynamics. It was named the Zeroth Law since it should have preceded the First and the Second Laws of Thermodynamics.

The First Law of Thermodynamics

The First Law of Thermodynamics

Statement: The total energy of an isolated system is constant; energy can be transformed from one form to another, but can be neither created nor destroyed.

Consider a system undergoing a series of adiabatic processes from a specified state 1 to another specified state 2. Being adiabatic, these processes obviously cannot involve any heat transfer, but they may involve several kinds of work interactions. Considering that there are an infinite number of ways to perform work interactions under adiabatic conditions, this statement appears to be very powerful, with a potential for far-reaching implications. This statement, which is largely based on the experiments of Joule in the first half of the 19th century, cannot be drawn from any other known physical principle and is recognized as a fundamental principle. This principle is called the First Law of Thermodynamics. A major consequence of the First Law is the existence and the definition of the property total energy E.

Considering that the net work is the same for all adiabatic processes of a closed system between two specified states, the value of the net work must depend on the end states of the system only, and thus it must correspond to a change in a property of the system. This property is the total energy. Note that the First Law makes no reference to the value of the total energy of a closed system at a state. It simply states that the change in the total energy during an adiabatic process must be equal to the net work done. Therefore, any convenient arbitrary value can be assigned to total energy at a specified state to serve as a reference point.

The Second Law of Thermodynamics

The Second Law of Thermodynamics

The Kelvin-Planck Statement: It is impossible for any device that operates on a cycle to receive heat from a single reservoir and produce net amount of work.

The Clausius Statement: It is impossible to construct a device that operates in a cycle, and produces no effect other than the transfer of heat from a lower temperature body to a higher temperature body.

It is common experience that a cup of hot coffee left in a cooler room eventually cools off. This process satisfies the First Law of Thermodynamics since the amount of energy lost by the coffee is equal to the amount gained by the surrounding air. Now consider the reverse process—the hot coffee getting even hotter in a cooler room as a result of heat transfer from the room air, it is obvious that this process never takes place.

As another familiar example, consider the heating of a room by the passage of electric current through a resistor. Again, the First Law dictates that the amount of electric energy supplied to the resistance wires be equal to the amount of energy transferred to the room air as heat. Now reverse this process. It will come as no surprise that transferring some heat to the wires does not cause an equivalent amount of electric energy to be generated in the wires(Fig. 1-3).

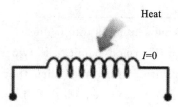

Figure 1-3 **Transferring heat to a wire will not generate electricity**

It is clear from these arguments that processes proceed in a certain direction and not in the reverse direction. The First Law places no restriction on the direction of a process, but satisfying the First Law does not ensure that the process can actually occur. This inadequacy of the First Law to identify whether a process can take place is remedied by introducing another general principle, the Second Law of Thermodynamics.

The use of the Second Law of Thermodynamics is not limited to identifying the direction of processes. The Second Law also asserts that energy has *quality* as well as *quantity*. The First Law is concerned with the quantity of energy and the transformations of energy from one form to another with no regard to its quality. Preserving the quality of energy is a major concern to engineers, and the Second Law provides the necessary means to determine the quality as well as the degree of degradation of energy during a process. As discussed later in this chapter, more of high-temperature energy can be converted to work, and thus it has a higher quality than the same amount of energy at a lower temperature.

The Second Law of Thermodynamics is also used in determining the theoretical limits for the performance of commonly used engineering systems, such as heat engines and refrigerators, as well as predicting the degree of completion of chemical reactions. The Second Law is also closely associated with the concept of perfection. In fact, the Second Law defines perfection for thermodynamic processes. It can be used to quantify the level of perfection of a process, and point the direction to eliminate imperfections effectively.

》》》 **Post-Reading Exercises**

1. Fill in the blanks.

(1) The First Law of Thermodynamics is recognized as the fundamental principle because _____.

(2) The Second Law of Thermodynamics is used in _____ and _____.

(3) The commonly used mercury-in-glass thermometer operates on the principle of _____.

2. Translate the following sentences into Chinese.

(1) The Zeroth Law of Thermodynamics: If two thermodynamic systems are each in thermal equilibrium with a third one, then they are in thermal equilibrium with each other. Accordingly, thermal equilibrium between systems is a transitive relation.

(2) The First Law of Thermodynamics: The total energy of an isolated system is constant; energy can be transformed from one form to another, but can be neither created nor destroyed.

(3) Kelvin-Planck Statement of the Second Law of Thermodynamics: It is impossible for any device that operates on a cycle to receive heat from a single reservoir and produce net amount of work.

(4) Clausius Statement of the Second Law of Thermodynamics: It is impossible to construct a device that operates in a cycle, and produces no effect other than the transfer of heat from a lower temperature body to a higher temperature body.

Unit 2　Fluid Mechanics

Fluid Mechanics

Warm-up Activities

1. Learn the following words and find relevant information.

            Chinese   Definition
   (1) pressure
   (2) density
   (3) viscosity
   (4) thermal conductivity
   (5) specific heat
   (6) compressible fluid
   (7) inviscid fluid
   (8) momentum
   (9) dimensionless number
   (10) internal(external) flow
   (11) ideal gas
   (12) buoyant force
   (13) acceleration of gravity
   (14) atmospheric pressure
   (15) coefficient of viscosity

2. Oral Exercise: Describe the following diagrams.

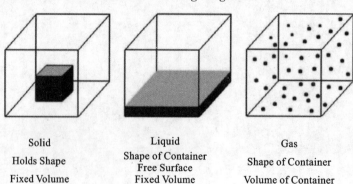

Solid　　　　　　　Liquid　　　　　　　Gas
Holds Shape　Shape of Container　Shape of Container
　　　　　　　Free Surface
Fixed Volume　Fixed Volume　Volume of Container

Intensive Reading

Fluid Mechanics

Mechanics is the oldest physical science that deals with both stationary and moving bodies under the influence of forces. The branch of mechanics that deals with bodies at rest is called statics, while the branch that deals with bodies in motion is called dynamics. The subcategory fluid mechanics is defined as the science that deals with the behaviour of fluids at rest (fluid statics) or in motion (fluid dynamics), and the interaction of fluids with solids or other fluids at the boundaries.

Fluid mechanics is also referred to as fluid dynamics by considering fluids at rest as a special case of motion with zero velocity. Fluid mechanics itself is also divided into several categories. The study of the motion of fluids that can be approximated as incompressible (such as liquids, especially water, and gases at low speeds) is usually referred to as hydrodynamics[1]. A subcategory of hydrodynamics is hydraulics, which deals with liquid flows in pipes and open channels. Gas dynamics deals with the flow of fluids that undergo significant density changes, such as the flow of gases through nozzles at high speeds. The category aerodynamics deals with the flow of gases (especially air) over bodies such as aircraft, rockets, and automobiles at high or low speeds. Some other specialized categories such as meteorology, oceanography, and hydrology deal with naturally occurring flows.

A substance in the liquid or gas phase is referred to as a fluid. Distinction between a solid and a fluid is made on the basis of the substance's ability to resist an applied shear (or tangential) stress that tends to change its shape. A solid can resist an applied shear stress by deforming, whereas a fluid deforms continuously under the influence of a shear stress, no matter how small. In solids, stress is proportional to strain, but in fluids, stress is proportional to strain rate. When a constant shear force is applied, a solid eventually stops deforming at some fixed strain angle, whereas a fluid never stops deforming and approaches a constant rate of strain. For a fluid at rest its shear stress is zero. When the walls are removed or a liquid container is tilted, a shear develops as the liquid moves to reestablish a horizontal free surface. In a liquid, groups of molecules can move relative to each other, but the volume remains relatively constant because of the strong cohesive forces between the molecules. As a result, a liquid takes the shape of the container it is in, and it forms a free surface in a larger container in a gravitational field. A gas, on the other hand, expands until it encounters the walls of the container and fills the entire available space. This is because the gas molecules are widely spaced, and the cohesive forces

fluid mechanics
流体力学

fluid statics
流体静力学
fluid dynamics
流体动力学

shear stress
剪应力

between them are very small. Unlike liquids, a gas in an open container cannot form a free surface. Although solids and fluids are easily distinguished in most cases, this distinction is not so clear in some borderline cases. For example, asphalt appears and behaves as a solid since it resists shear stress for short periods of time. When these forces are exerted over extended periods of time, however, the asphalt deforms slowly, behaving as a fluid. Some plastics, lead, and slurry mixtures exhibit similar behaviour.

Fluid flow is often confined by solid surfaces, and it is important to understand how the presence of solid surfaces affects fluid flow. Just like water in a river cannot flow through large rocks, and must go around them. That is because water velocity normal to the rock surface must be zero, and water approaching the surface normally comes to a complete stop at the surface. What is not as obvious is that water approaching the rock at any angle also comes to a complete stop at the rock surface, and thus the tangential velocity of water at the surface is also zero. Consider the flow of a fluid in a stationary pipe or over a solid surface that is nonporous. All experimental observations indicate that a fluid in motion comes to a complete stop at the surface and assumes a zero-velocity relative to the surface. That is, a fluid in direct contact with a solid "sticks" to the surface, and there is no slip[2]. This is known as the *no-slip condition* (Fig. 2-1). The fluid property responsible for the no-slip condition and the development of the boundary layer is viscosity.

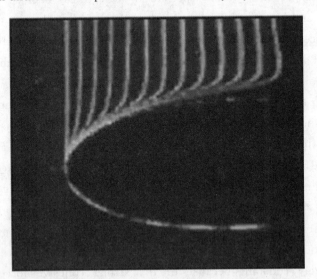

Figure 2-1 The development of a velocity profile due to the no-slip condition as a fluid flows over a blunt nose

Classification of Fluid Flow

There is a wide variety of fluid flow problems encountered in practice, and it is usually convenient to classify them on the basis of some common characteristics to make it feasible to study them in groups [3]. There are many ways to classify fluid flow problems, and here some general categories are presented.

Viscous vs Inviscid Regions of Flow

When two fluid layers move relative to each other, a friction force develops between them and the slower layer tries to slow down the faster layer. This internal resistance to flow is quantified by the fluid property viscosity, which is a measure of internal stickiness of the fluid. Viscosity is caused by cohesive forces between the molecules in liquids and by molecular collisions in gases. There is no fluid with zero viscosity, and thus all fluid flows involve viscous effects to some degree. Flows in which the frictional effects are significant are called viscous flows.

However, in many flows of practical interest, there are regions (typically regions not close to solid surfaces) where viscous forces are negligibly small compared to inertial or pressure forces. Neglecting the viscous terms in such inviscid flow regions greatly simplifies the analysis without much loss in accuracy [4]. The development of viscous and inviscid regions of flow as a result of inserting a flat plate parallel into a fluid stream of uniform velocity. The fluid sticks to the plate on both sides because of the no-slip condition, and the thin boundary layer in which the viscous effects are significant near the plate surface is the viscous flow region. The region of flow on both sides away from the plate and largely unaffected by the presence of the plate is the inviscid flow region.

Laminar vs Turbulent Flow

Some flows are smooth and orderly while others are rather chaotic. The highly ordered fluid motion characterized by smooth layers of fluid is called laminar. The word laminar comes from the movement of adjacent fluid particles together in "laminae". The flow of high-viscosity fluids such as oils at low velocities is typically laminar. The highly disordered fluid motion that typically occurs at high velocities and is characterized by velocity fluctuations is called turbulent. The flow of low-viscosity fluids such as air at high velocities is typically turbulent. A flow that alternates between being laminar and turbulent is called transitional. The experiments conducted by Osborne Reynolds in the 1880s resulted in the establishment of the dimensionless Reynolds number, as the key parameter for the determination of the flow regime in pipes.

Steady vs Unsteady Flow

The terms steady and uniform are used frequently in engineering, and thus it is important to have a clear understanding of their meanings. The term steady implies no change of properties, velocity, temperature, etc., at a point with time. The

viscous
黏性的
inviscid
无黏性的
friction
摩擦

parallel
平行的

laminar
层流
turbulent
湍流

steady
稳定的
unsteady
不稳定的

opposite of steady is unsteady. The term uniform implies no change with location over a specified region. These meanings are consistent with their everyday use. The terms unsteady and transient are often used interchangeably, but these terms are not synonyms. In fluid mechanics, unsteady is the most general term that applies to any flow that is not steady, but transient is typically used for developing flows. When a rocket engine is fired up, for example, there are transient effects (the pressure builds up inside the rocket engine, the flow accelerates, etc.) until the engine settles down and operates steadily. The term periodic refers to the kind of unsteady flow in which the flow oscillates about a steady mean. Many devices such as turbines, compressors, boilers, condensers, and heat exchangers operate for long periods of time under the same conditions, and they are classified as steady-flow devices. During steady flow, the fluid properties can change from point to point within a device, but at any fixed point they remain constant [5]. Therefore, the volume, the mass, and the total energy content of a steady-flow device or flow section remain constant in steady operation.

synonyms
同义词

Notes

[1]研究近似不可压缩流体(如液体,特别是水和低速气体)运动的学科通常称为流体力学。

[2]所有的实验观测都表明,运动中的流体在表面完全停止,并假定相对于表面的速度为零。也就是说,液体"粘附"在与它直接接触的固体表面,没有滑动。

[3]实践中遇到的流体流动问题种类繁多,为方便起见,通常根据一些共同的特点对其进行分类以便分组研究。

[4]忽略这类无黏性流体流动中的黏性,可以大大简化分析,且不会造成很大的精度损失。

[5]在稳定流动过程中,流体特性可以在装置内的各个点之间变化,但在任何固定点,它们都保持不变。

Extensive Reading

Heat Exchanger

Heat exchangers are devices that facilitate the exchange of heat between two fluids that are at different temperatures while keeping them from mixing with each other. Heat exchangers are commonly used in practice in a wide range of applications, from heating and air-conditioning systems in a household to chemical processing and power production in large plants. Heat exchangers differ from mixing chambers in that they do not allow the two fluids involved to mix. Heat transfer in a heat exchanger usually involves convection in each fluid and conduction through the wall separating the two fluids.

Types of Heat Exchangers

Different heat transfer applications require different types of hardware and different configurations of heat transfer equipment. The attempt to match the heat transfer hardware to the heat transfer requirements within the specified constraints has resulted in numerous types of innovative heat exchanger designs.

The simplest type of heat exchanger consists of two concentric pipes of different diameters, as shown in Fig. 2-2, called the double-pipe heat exchanger. One fluid in a double-pipe heat exchanger flows through the smaller pipe while the other fluid flows through the annular space between the two pipes. Two types of flow arrangement are possible in a double-pipe heat exchanger: in *parallel flow*, both the hot and cold fluids enter the heat exchanger at the same end and move in the same direction. In *counter flow*, on the other hand, the hot and cold fluids enter the heat exchanger at opposite ends and flow in opposite directions. Another type of heat exchanger, which is specifically designed to realize a large heat transfer surface area per unit volume, is the *compact* heat exchanger. The flow passages in these compact heat exchangers are usually small and the flow can be considered to be laminar. Compact heat exchangers enable us to achieve high heat transfer rates between two fluids in a small volume, and they are commonly used in applications with strict limitations on the weight and volume of heat exchangers.

In compact heat exchangers, the two fluids usually move perpendicular to each other, and such flow configuration is called *cross-flow*. The cross-flow is further classified as unmixed and mixed flow, depending on the flow configuration, as shown in Fig. 2-2. In (c) the cross-flow is said to be unmixed since the plate fins force the fluid to flow through a particular inter fin spacing and prevent it from moving in the transverse direction (i.e., parallel to the tubes). The cross-flow in (d) is said to be mixed since the fluid now is free to move in the transverse direction. Both fluids are unmixed in a car radiator. The presence of mixing in the fluid can have a significant effect on the heat transfer characteristics of the heat exchanger.

A heat exchanger typically involves two flowing fluids separated by a solid wall. Heat is first transferred from the hot fluid to the wall by convection, through the wall by conduction, and from the wall to the cold fluid again by convection. Any radiation effects are usually included in the convection heat transfer coefficients.

In the analysis of heat exchangers, it is convenient to work with an overall heat transfer coefficient U that accounts for the contribution of all these effects on heat transfer, and to express the rate of heat transfer between the two fluids as

$$Q = \frac{\Delta T}{R} = U A_s \Delta T \tag{1}$$

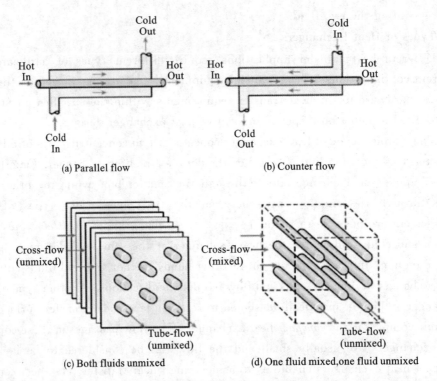

Figure 2-2　Different flow configurations

where A_s is the surface area and U is the overall heat transfer coefficient, whose unit is W/(m² · K).

When the wall thickness of the tube is small and the thermal conductivity of the tube material is high, as is usually the case, the thermal resistance of the tube is negligible and the inner and outer surfaces of the tube are almost identical. Then overall heat transfer coefficient U can be simplified to

$$\frac{1}{U}=\frac{1}{h_i}+\frac{1}{h_0} \tag{2}$$

The rate of heat transfer between the two fluids at a location in a heat exchanger depends on the magnitude of the temperature difference at that location, which varies along the heat exchanger.

Fouling Factor

The performance of heat exchangers usually deteriorates with time as a result of accumulation of deposits on heat transfer surfaces. The layer of deposits represents additional resistance to heat transfer and causes the rate of heat transfer in a heat exchanger to decrease. The net effect of these accumulations on heat transfer is represented by a fouling factor, which is a measure of the thermal resistance introduced by fouling.

The most common type of fouling is the precipitation of solid deposits in a fluid on the heat transfer surfaces. You can observe this type of fouling even in your house. If you check the inner surfaces of your teapot after prolonged use, you will probably notice a layer of calcium-based deposits on the surfaces at which boiling occurs. This is especially the case in areas where the water is hard. The scales of such deposits come off by scratching, and the surfaces of such deposits can be cleaned by chemical treatment. Now imagine those mineral deposits forming on the inner surfaces of fine tubes in a heat exchanger (Fig. 2-3) and the detrimental effect it may have on the flow passage area and the heat transfer. To avoid this potential problem, water in power and process plants is extensively treated and its solid contents are removed before it is allowed to circulate through the system. The solid ash particles in the flue gases accumulating on the surfaces of air preheaters create similar problems.

Figure 2-3 Precipitation fouling of ash particles on superheater tubes

Another form of fouling, which is common in the chemical process industry, is corrosion and other chemical fouling. In this case, the surfaces are fouled by the accumulation of the products of chemical reactions on the surfaces. This form of fouling can be avoided by coating metal pipes with glass or using plastic pipes instead of metal ones. Heat exchangers may also be fouled by the growth of algae in warm fluids. This type of fouling is called biological fouling and can be prevented by chemical treatment. In applications where it is likely to occur, fouling should be considered in the design and selection of heat exchangers. In such applications, it may be necessary to select a larger and thus more expensive heat exchanger to ensure that it meets the design heat transfer requirements even after fouling occurs. The periodic cleaning of heat exchangers and the resulting down time are additional penalties associated with fouling.

▶▶▶ Post-Reading Exercises

1. Answer the following questions.

(1) What are the factors the rate of heat transfer in a heat exchanger depends on?

(2) What are the differences between double-pipe heat exchangers and compact ones?

(3) What are the causes of fouling?

2. Translate the following sentences into Chinese.

(1) In the analysis of heat exchangers, it is convenient to work with an overall heat transfer coefficient U that accounts for the contribution of all these effects on heat transfer.

(2) Now imagine those mineral deposits forming on the inner surfaces of fine tubes in a heat exchanger and the detrimental effect it may have on the flow passage area and the heat transfer.

(3) In applications where it is likely to occur, fouling should be considered in the design and selection of heat exchangers. In such applications, it may be necessary to select a larger and thus more expensive heat exchanger to ensure that it meets the design heat transfer requirements even after fouling occurs.

Computational Fluid Dynamics

Computational fluid dynamics (CFD) is a technique used __1__ model the behaviour of fluids. In building design, it is typically used to model the movement and temperature of air __2__ spaces. This is important as it __3__ designers to investigate internal conditions before a building is built, allowing them to test options and select the most effective solutions. CFD can also be used to investigate the impact of a new building on air movement around a site, and has been used to model other "fluid" behaviour, such as the movement of people.

Simulations are typically run __4__ a number of different scenarios, __5__ behaviour under different levels of occupancy, different climatic conditions, in different modes of building services operation, with different openings between spaces and so on. This can build __6__ an overall picture of how a building is likely to behave under normal operating conditions as well as during unusual or extreme conditions.

How It Works

CFD works by dividing a space into a grid containing a large number of "cells". The grid of cells is surrounded by boundaries that simulate the surfaces and openings that enclose the space. The temperature of the boundaries, the air movement at openings, and the air temperature within the cells are then set to a starting condition which it is hoped is close to those that might be expected to be found within the space. These conditions might be determined using a boundary model that predicts boundary conditions, __7__ climatic and materials data. The more accurate the boundary model is, and the closer the starting condition is to the __8__ position predicted by the model, the faster the model will run and the more accurate the output is likely to be.

The software will then simulate the flow of air from each cell to those surrounding it, and the exchange of heat between the boundary surfaces and the cells adjacent to them. After a series of iterations, the model will come to a steady state that represents the actual air velocities and distribution of temperatures expected to be found within the space.

CFD can be a very useful tool in the right hands, and the output graphics are very __9__ and seductive. However results are highly dependent on the knowledge of the person __10__ the results. This is an increasing concern as CFD software becomes more straight-forward to use and so is more easily operated by people with little understanding of the mathematical model that underpins it.

If the input information is wrong, the output information will be as well. CFD is no substitute for common sense. An important consideration in developing a CFD model is the generation of the grid of cells. The greater the number of cells are, the more accurate the simulation will be, but the longer the model will take to run. In some parts of a space, using a large cell size may not have a significant impact on the results, however in sensitive areas, for example around complex boundaries or where there might be a large temperature difference between a boundary surface and the air next to it, it is important that cells are as small as is computationally practical. For example, a very small cell size (a fine grid mesh) is necessary to properly simulate the downdraft that can be experienced next to a cold window. If such a downdraft is not simulated, the heat exchange between the window and the space it encloses will be underestimated.

In spaces where the surfaces enclosing the space are non-cartesian (ie they are curved, or at an angle rather than purely horizontal or vertical) it is important that the grid is body-fitted (ie that it follows the contours of the surfaces) rather than cartesian (in a series of steps), otherwise air velocity at the surface will be underestimated and so heat transfer between the space and its enclosing surfaces will be underestimated. This is particularly important where there is a large temperature difference between the surface and the air adjacent to it.

Where CFD is being used to predict user comfort within a space, it is important that both air temperature and radiant temperature are considered. CFD in itself only models air temperature and air velocity, however, around half of the contribution to our thermal comfort within buildings is dependent on radiant heat transfer, ie the temperature of the surfaces. Some CFD software is able to include radiant influences on the temperatures that will be felt by occupants.

Post-Reading Exercises

1. Cloze.

	A	B	C	D
1 ( )	to	for	as	of
2 ( )	to	from	in	inside
3 ( )	allow	allows	permit	permits
4 ( )	with	as	use	in
5 ( )	model	modelling	modelled	to model
6 ( )	to	for	as	of
7 ( )	based on	depend on	due to	as
8 ( )	finish	final	ending	closing
9 ( )	persuasive	persuade	persuading	persuasion
10 ( )	interpret	to interpret	interpreting	interpretation

2. True or False.

(1) CFD can be substitute for common sense. ( )

(2) The greater the number of cells is, the more accurate the simulation will be. (　)

(3) Grid shall follow the contours of the surfaces. (　)

(4) Temperature and radiant temperature are important for human comfort. (　)

(5) CFD helps designers to test their design. (　)

Unit 3 Built Environment

Warm-up Activities

1. Learn the following words and make out their Chinese names.

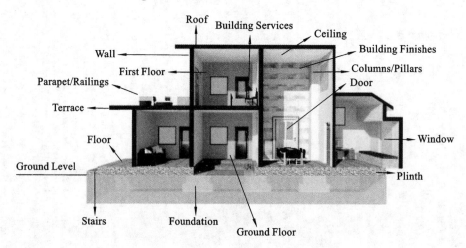

2. Match the words to right description.

(1) Roof　　　　　• A structural unit that uniformly distributes the load from the superstructure to the underlying soil

(2) Beams　　　　• Horizontal members used to support vertical applied loads across an opening

(3) Columns　　　• It forms the topmost component of a building structure

(4) Floor　　　　 • Vertical members constructed above the ground level

(5) Foundation　　• Surface laid on the plinth level

(6) Walls　　　　 • A sequence of steps that connects different floors in a building structure

(7) Stairs　　　　• Vertical elements which support the roof and can be made from stones, bricks, concrete blocks, etc

Intensive Reading

Building Thermal Environment

The Construction Industry Council (CIC) suggests that the built environment... encompass all forms of building (housing, industrial, commercial, hospitals, schools, etc), and civil engineering infrastructure, both above and below ground and includes the managed landscapes between and around building [1]. It is essentially an enclosure within which activities can be carried out. It is a structure, usually consisting of a roof, walls, floors and openings such as doors and windows that is generally positioned permanently in one location [2].

Thermal Environment

The primary effect of the environment on occupants is that of ambient conditions on heat exchange. At a given point in time and space, an individual who occupies that point experiences a specific set of physical conditions. Those conditions that determine the potential for heat exchange between the individual and the surrounding environment define the thermal environment [3]. Within the built environment, the thermal environment can be influenced by:

- Passive building design (shading, insulation, thermal mass, etc).
- Active building systems (heating, cooling and air conditioning).
- Personal behaviour (removing clothing, activity level, etc).

Human Comfort

One of people's basic needs is to maintain a constant body temperature while the metabolism regulates heat flows from the body to compensate for changes in the environment [4].

Heat transfer between the human body and its surroundings takes place through conduction, convection and radiation. Points of conduction contact with the structure are made with furniture and floor. Clothing normally has a substantial thermal insulation value and discomfort should be avoided. Heat removed from the body by natural convection currents in the room air, or fast-moving air streams produced by ventilation fans or external wind pressure is a major source of cooling. The body's response to a cool air environment is to restrict blood circulation to the skin to conserve deep tissue perpetrator, involuntary reflex action, shivering, if necessary, and in extreme cases, inevitable lowering of body temperature. This last state of hypothermia can lead to loss of life and is a particular concern in relation to elderly people. Radiation heat transfer takes place between the body and its surroundings. The direction of heat transfer may be either way, but normally a

---

built environment 建筑环境

infrastructure 基础设施

landscape 风景

heat exchange 热交换

passive building design 被动建筑设计

active building design 主动建筑设计

human comfort 人体舒适度

minor part of the total body heat loss takes place by this method. Radiation between skin and clothing surfaces and the room depends on the fourth power of the absolute surface temperature, the emissivity, the surface area and the geometric configuration of the emitting. Thus, a moving person will experience changes in comfort level depending on the location of the hot and cold surfaces in the room, even though air temperature and speed may be constant. Some source of radiant heat is essential for comfort, particularly for sedentary occupations. Hot water central heating radiators, direct fuel-fired appliances and most electrical heater provide a combination of convection and radiated warmth. The elderly find particular difficulty in keeping warm when they are relatively immobile, and convective heating alone is unlikely to be satisfactory. A source of radiant heat provides rapid heat transfer and a focal point, easy manual control and quick heat-up periods. Overheating from sunshine causes discomfort and disability glare. Humid air is exhaled; further transfer of moisture from the body takes place by evaporation from the skin and through clothing. Maintenance of a steady rate of moisture removal from the body is essential; this is a mass transfer process depending on air humidity, temperature and speed as well as variables such as clothing and activity.

Outside air ventilation rates vary for different applications: 10 L/s per person is generally used for fume removal. Measurement of return air CO_2 content can be applied to ensure adequate air quality. The designer needs to evaluate all aspects of the need for outdoor air ventilation. These include the heating and cooling plant loads that will be generated, the potential for energy recovery between the incoming fresh air and outgoing air at room temperature, avoidance of draughts in the occupied rooms, the variation of load with the occupancy level and the ability to utilize outdoor air to provide free cooling to the building when the outdoor air is between 10℃ and 20℃. Moving-air velocity in normally occupied rooms will be between 0 and 2 m/s where the upper figure relates to a significantly uncomfortable hot or cold draught. Still-air conditions are most unlikely to occur, all buildings leak air convection currents from people, warmed surfaces and electrical equipment and computers promote air circulation. Room air movement patterns should be variable rather than monotonous and ventilation of every part of the space is important.

Mean radiant temperature is measure of radiation heat transfers taking place between various surfaces and has an important bearing on thermal comfort. Air temperature is that from a mercury-in-glass dry-bulb thermometer freely exposed to the air stream, usually in sling psychrometer. Atmospheric vapour pressure is that part of the barometric pressure produced by the water vapour in humid air. Standard atmospheric pressure at sea level is 1013.25 mb comprising about 993.0 mb from the weight of dry gases and 20.25 mb from the water vapour, depending on the values of barometric pressure, air temperature and humidity.

The main comfort criteria for sedentary occupants in buildings in climates similar to that of the British Isles are: operative temperature 19-23℃ depending on room use, feeling of freshness, mean radiant temperature is slightly above air temperature. Air temperature and the mean radiant temperature should be approximately the same, as large differences cause either radiant overheating or excessive heat loss from the body to the environment, as would be experienced during occupation of a glasshouse through seasonal variations. Warmer temperatures become acceptable in summer where an upper limit of 26℃ db. can be tolerated for short periods. Percentage saturation should be in the range of 40-70%. Maximum air velocity at the neck should be 0.1 m/s for a moving-air temperature of 20℃ db. Hot and cold draughts should be avoided. Variable air velocity and direction are preferable to constant and this is achieved by changes in natural ventilation from prevailing wind, movement of people around the building, on-off or high-low thermostatic operation of fan-assisted heaters or variable-volume air conditioning systems. The minimum quantity of fresh air for room use that will remove probable contamination is 10 L/s per person. Mechanical ventilation systems should provide at least four air changes per hour to avoid stagnant pockets and ensure good air circulation. Incoming fresh air can be filtered to maintain a clean, dust-free, internal environment. The difference between room air temperatures at head and foot levels should be no more than 1℃. Ventilation air quantity can be determined by some other controlling parameter, for example, removal of smoke, fumes or dust, solar or other heat gains and dilution of noxious fumes. Living in tropical climates requires adaptation to constant high humidity and 30℃ db. outdoors; indoor humidity of 40% and above in arid climates means that it is raining.

Environmental Measurements

Measurement of air temperature and humidity is accurately made by using both dry-and wet-bulb mercury-in-glass thermometers in a sling psychrometer, otherwise called a whirling hygrometer. The dry-bulb thermometer defines the air temperature ℃ db. and evaporation of water from the cotton wick cools the wet-bulb thermometer; its reading is known as the wet-bulb temperature ℃ wb.

In order to find the mean radiant temperature t_r, the dry-bulb temperature t_a and the air velocity v m/s are measured:

$$t_r = \tan(1+2.35\sqrt{v}) - 2.35 t_a \sqrt{v} \tag{1}$$

where t_a is the globe temperature measured at the center of a blackened globe of diameter 150 mm suspended at the measurement location. In normally occupied

operative temperature
工作温度

draught
气流
prevailing wind
盛行风

wet-bulb temperature
湿球温度
whirling hygrometer
旋转湿度计

blackened globe
发黑的球体

rooms, the air temperature and the mean radiant temperature should be within a few degrees of each other.

Direct-reading air velocity measuring instruments are shown in Fig. 3-1 and Fig. 3-2. Rotating vane anemometers are used to measure air streams through ventilation grilles, where the rotational speed of the blades is magnetically counted [5]. Thermistor and hot-wire anemometers use the air stream cooling effect on the probe. Duct air flow rates are found by inserting a pitot-static tube into the airway, taking up to 48 velocity readings and evaluating the air volume flow rate from the average air velocity found from all the readings [6].

rotating vane anemometers
旋转叶片风速计

Figure 3-1　Rotating vane anemometers　　　**Figure 3-2　Hot-wire anemometers**

The term of air humidity is percentage saturation, and the most reliable method of measurement is to take dry and wet bulb air temperature readings using a sling psychrometer and refer to a psychrometric chart.

Notes

[1]建筑业委员会(CIC)认为,建筑环境……包括所有地上和地下的各种形式的建筑(住房、工业、商业、医院、学校等)和土木工程基础设施,以及建筑物之间和周围的景观。

[2](建筑)本质上是一个可以进行活动的场所。它通常是一种由屋顶、墙壁、地板和开口(如门和窗)组成的结构,通常固定在一个位置。

[3]决定个体与周围环境之间热交换潜力的条件定义了热环境。

[4]人们的基本需求之一是保持恒定的体温,同时新陈代谢调节来自身体的热量以补偿环境的变化。

[5]旋转叶片风速计通过通风格栅测量气流,其中叶片的转速通过磁力计测量。

[6]通过将皮托静压管插入气道,最多获取 48 个速度读数,并根据从所有读数中获得的平均空气速度来评估空气体积流率,从而确定管道空气流速。

Extensive Reading

Energy and Building

Energy Consumption

The world final energy consumption amounts to 4364 million tons of standard coal in the year 2016. Hereof, 20.6%, falls to households and small trade, mainly for heating and cooling buildings. A large share of the worldwide energy consumption is caused by households. This embodies a vast potential for energy savings, because an optimisation of energy use for households has not yet taken place. The reason for this is the long service life or building lifetime, respectively. This period is much longer than for other goods, especially in technology. A building is technologically outdated at the earliest after 30 years. This leads to a comparatively long time for the distribution of newly established energy efficient standards by new buildings. Through additional measures for an energy retrofit in conjunction with renovation of the building stock, a huge potential results for the aspired saving of resources.

Heat Balance of Buildings

The heat balance of a building includes all sources and sinks of energy inside a building, as well as all energy flows through its envelope. This envelope encloses the volume which is kept above a set temperature for all weather conditions by the use of heating energy. The extend of all heat flows, which do hereby occur, is either dependent on external or internal influence factors (weather, user). These heat flows can be arranged into five categories:

(1) **Transmission losses** L_T are those amounts of heat, which flow through the building envelope from inside to outside by conduction or heat transfer, respectively.

(2) **Ventilation losses** L_V are caused by exchange of warm indoor air with colder outdoor air. The user independent air exchange is through joints by infiltration or exfiltration, respectively. In addition, room air can be exchanged through open windows or by a mechanical ventilation system. Ventilation is indispensable, up to a certain extend, to assure the hygienically necessary air exchange rate.

(3) **Solar gains** G_S are irradiations of solar energy through windows and other transparent or translucent constructional elements. Also added to the solar gains, is that part of the solar heating of the opaque building envelope, from which the indoor

area benefits.

(4) **Internal gains** G_I are heat outputs from persons, appliances, computers and other electric devices, as well as from illumination.

(5) **Heating demand** H is exactly that amount of energy, which is necessary to maintain the desired room temperature by compensating the excess of losses compared to the gains.

The gains and losses are specified for a certain period of time. Division of this value by the corresponding area of heated space in m², gives all heat flows (1 to 5) in the usual unit for the (floor space) specific energy demand for heating: kWh/(m² · a). The allocation of the transmission and ventilation to gains or losses depends, strictly speaking, on whether the outdoor temperature is higher or lower than the room temperature. If the gains exceed the losses for a longer period of time, the desired indoor temperature would be overstepped. Instead of heating, cooling would be necessary for the balance. This case occurs in summer and is treated separately.

However, it can happen in winter, but especially in the intermediate times (autumn, spring), that the set temperature is sporadically exceeded by high solar or internal gains. The total monthly internal and solar gains are, however, not to 100% effective for heating. Therefore, they are rated with a utilisation factor $F_U <$ 1. Typical values for the yearly mean value are in the range of 0.5-0.9, depending on the heating energy demand and the kind of construction. For an extended period of time the changes of the stored energy in the building mass, indicated by the mean building temperature, are negligible, and the energy balance is:

$$H = L_T + L_V - F_U \cdot (G_S + G_I) \tag{2}$$

This means, the heating energy corresponds to the sum of the losses, reduced by the utilised part of the gains.

Factors of Influence

The heating energy demand of a building is influenced by the climate conditions at the location, the immediate building surroundings, and the behaviour of the users.

The most important climatic influences are outdoor temperature and solar radiation. The heating energy demand for a heating period depends on the level of the outdoor temperature as well as the length of this period. Both influence factors are described by the heating degree days (HDD). This is based on an indoor temperature of 20℃ and on a heating temperature of 15℃, i.e. heating is necessary only on days, where the outdoor temperature is less than 15℃. For every day, on which this condition is fulfilled, the difference between 20℃ and the mean outdoor temperature is summed. This gives the heating degree days on the basis (20/15), denoted HDD_{15}. For buildings with high insulation level (low energy buildings) a heating temperature of 12 ℃ is assumed. This gives the heating degree days HDD_{12} on the basis (20/12). The transmission and ventilation losses are directly proportional to the heating degree days, which can vary significantly for different locations (and also for different years). The losses for a certain building depend on the location and the corresponding heating degree days. Also, a variation of the desired indoor temperature leads to a change of the losses, which is described by the related heating degree days. For 200 heating days per year and an increase or decrease of the indoor temperature by 1℃, the heating degree days change 200 Kd in both cases.

A further influence factor for the heating energy demand is the solar radiation. The higher the irradiation at the location of the building, the higher the solar gains can be. The influence is, however, much smaller than that of the indoor and outdoor temperatures. At buildings with high glazing to wall ratio (more than 40%) or large areas of transparent insulation, the influence of solar radiation can become significant. For the dependence of the solar gains on the global radiation, the individual shading situation, e. g. by neighbouring buildings, the typography and vegetation plays an important role.

Finally, the influence of the wind on the heat losses is noteworthy. The pressure distribution on the facade, generated from the wind velocity and its direction, depends on the character of the immediate surroundings. It effects the total ventilation losses of a building by infiltration and exfiltration through joints in the building envelope. For modern buildings, which are largely wind- and airtight, this influence is, however, inferior.

In addition to the above-mentioned set temperature, the building user influences the heating energy demand of a building by his ventilation behaviour. Inadequately strong window ventilation can push the heating energy demand to the limit of the furnace performance. Even further ventilation losses lead to a sinking temperature level, resulting in discomfort. From the point of view of energy saving, several short (5 to 10 minutes) and strong ventilation occurrences with heating switched off, are more prosperous than longer lasting ventilation with slightly opened windows. For the accurate adjustment of the hygienically necessary air exchange rate, and, therefore, to optimise and limit the ventilation losses, a mechanical ventilation system is ingenious.

Building Services

The role of the building services is to cover the demand of the user for water, electricity, heat and increasingly also fresh air. The building services, relevant for energy, are mainly heating system, domestic hot water and ventilation system. The different building standards have discriminative requirements and possibilities with respect to the building services. In houses of the building stock, a powerful heating system is necessary to cover the heating demand and to make the heating comfortable. Here, usually a gas, oil or electric heating system is used, even in smaller buildings. In buildings with higher insulation standard (low energy houses), energy sources with lower energy density, like wood heating (e. g. pellets), can also be used. Moreover, the temperature of the heat transfer can be lower ($<$ 35 ℃), so that a supplemental solar collector are energetically favourable. In addition to the high insulation level in passive houses, a ventilation system with highly efficient heat recovery is utilized to save expenses for a conventional heating system and the corresponding infrastructure (oil tank, gas connection, furnace room, chimney, separate heat transfer system, heat exchanger). The low heating energy demand is covered by a small heating module, e. g. by liquid gas or electricity, or by an adequately dimensioned heat pump. As heat distributor, the ventilation system is used (which is installed anyway). The domestic hot water is usually provided in passive houses by a solar collector, which is supplemented by the existing heating system. The building services of self sufficient houses eventually have to supply also the (small) rest of the heating energy demand by additional technical effort. This can be achieved by seasonal repositories, which store the solar energy over collectors during the summer into a central tank, laid out often for several buildings (storage medium generally water). This energy is used as heating energy in winter. Hot water storage tanks can store 60 to 80 kWh thermal energy in each m^3 of water.

>>> **Post-Reading Exercises**

1. Based on Section "Heat Balance of Buildings", what are the elements of the heat balance?

2. Choose the best answer for each of the following.
(1) HDD_{16} means the heating temperature is _____ ℃.
   A. 12
   B. 16
   C. 20
(2) The transmission and ventilation losses are _____ proportional to the heating degree days.
   A. directly

  B. reversely
  C. exponentially
(3) Which of the factors does not influence the energy demand? _____.
  A. Outdoor temperature
  B. Internal temperature
  C. The users
(4) The heat balance of a building includes _____.
  A. all sources and sinks of energy inside a building
  B. solar energy
  C. ventilation heat gain
(5) The energy consumed by households is _____.
  A. 117 EJ
  B. 4364 million tons of standard coal
  C. 898 million tons of standard coal

3. Translation.

(1) The world final energy consumption amounts to 4364 million tons of standard coal in the year 2016. Hereof, 20.6%, falls to households and small trade, mainly for heating and cooling buildings.

(2) The pressure distribution on the facade, generated from the wind velocity and its direction, depends on the character of the immediate surroundings.

Nearly Zero Energy Buildings

  The energy demand related to buildings can be reduced by energy-saving technologies and energy-efficient regulation on the buildings. The increase in energy performance of the buildings leads to the concept of low energy building or Nearly Zero Energy Buildings (nZEB).

  Based on the definition of European Parliament, the nZEBs are expected to use renewable energy technologies (RETs) on-site or nearby to meet energy demands. The definition of EISA consists of (1) reducing the energy demands, (2) meeting the energy needs from the non-carbon emission energy generations, (3) increasing the practice related to nZGHG, and (4) minimizing the installation and running cost. According to Riedy et al. zero emission buildings can be defined as near zero energy, zero energy, passive house, 100% renewable, carbon neutral, climate positive and positive advancement, energy plus, and zero net energy.

Inside typical building systems' boundary, the building consumes delivered energy on-site such as electricity district heating/cooling, natural gas, biomass, other fuels, and finally renewable energy generation. Excessive electricity is also sent to the grids. The nZEB can be described as annual energy consumed by the building irrespective of the life cycle. The target of conventional nZEB provides the balance between loading and generating the energy within the buildings.

Fig. 3-3 presents three different energy-efficient techniques such as passive service system and renewable energy systems. Passive systems consist of building orientation, envelope, airtightness, and shade. When implementing the passive systems to buildings, thermal and electrical energy consumption decreases effectively. In addition, in order to offer comfortable temperatures to the buildings, HVAC, domestic hot water systems, and lighting indoors can be reinvented to reduce energy loads. In this way, the performance of building energy systems is increased through integration of RETs. RETs are used not only to generate electricity but also to heat and cool the indoor environment via combined heating and cooling and power solutions (tri-generation systems). Through nZEB, thermally comfortable living spaces can still be achieved for dwellers.

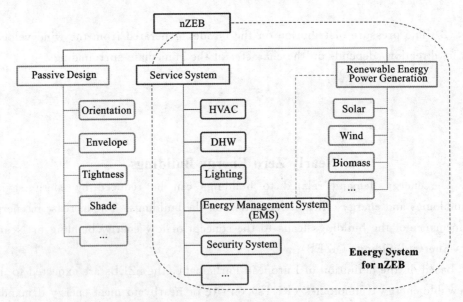

Figure 3-3 Design elements for nZEB

The retrofitting of the existing buildings toward nZEBs is really important than the newly constructed buildings. Since the energy-efficient materials for the new buildings are commercially available on market, the main challenge comes from the existing buildings. By looking into reports, the buildings that have existed from the 1960s in Europe are about 40% of all buildings in Europe nowadays. Newly constructed buildings in Europe are attributed to 1% of the building stock. It is predicted that the buildings existing in Europe today might be utilized until 2050. The energy performance of the existing buildings is relatively poor. Hence, the retrofits of these buildings are vital parts of the target of 2050. Along with improving energy performance of the building, the economic growth and the life quality also increase proportionally. It is widely believed that the application of retrofitting the existing buildings will comprise a wide range of developments including thermal insulation of building facade and roofs, upgrading the space heating and cooling systems, renovation of electrical and electronic appliances, and utilizing RETs on-site or nearby. Looking at the report presented by European Commissions, it is predicted that minimum energy saving can be targeted by 2020 and the amount saved would be in the range of 60-80 Mtoe/year. So, to attain the target of near zero carbon world in 2080, 43 billion buildings would need to be constructed or retrofitted based on the zero energy standards per year.

▶▶▶ Post-Reading Exercises

1. Choose the best answer for each of the following.
  (1) Who thinks nZEBs are expected to meet the energy needs from the non-carbon emission energy generations? _____.
     A. EISA
     B. European Parliament
     C. Riedy
  (2) Which is not a part of passive system? _____.
     A. Building orientation
     B. Building shape
     C. Shading
  (3) Which of the following is correct about newly constructed buildings in Europe? _____.
     A. They are about 40% of all buildings in Europe nowadays
     B. Newly constructed buildings in Europe are attributed to 40% of the building stock
     C. They might be utilized until 2050
  (4) Tri-generation system is not _____.
     A. combined heating and cooling and power solutions
     B. a part of RETs

  C. a system used to monitor energy consumption

2. Fill in the blanks with the words or expressions given below, and change the form where necessary.

  A zero-energy (ZE) building, also _____ as a zero net energy (ZNE) building, net-zero energy building (NZEB), net zero building is a building _____ zero net energy consumption, meaning the total amount of energy used by the building on an annual _____ is equal to the amount of renewable energy created _____, or in other definitions by renewable energy sources offsite. In some cases these buildings _____ contribute less overall greenhouse gas to the atmosphere during operations than similar non-ZNE buildings. They do at times consume non-renewable energy and produce greenhouse gases, but at other times reduce energy consumption and greenhouse gas production _____ by the same amount. A similar concept approved and _____ by the European Union and other _____ countries is nearly Zero Energy Building (nZEB), with the goal of having all buildings in the region under nZEB standards _____ 2020. Terminology tends to vary _____ countries and agencies; the IEA and European Union most commonly use "net zero", with "zero net" mainly used in the USA.

| elsewhere | agree | basis | by | consequent |
| between | onsite | with | implement | known |

3. Translation.

(1) Inside typical building systems boundary, the building consumes delivered energy on-site such as electricity district heating/cooling, natural gas, biomass, other fuels, and finally renewable energy generation.

(2) Since the energy-efficient materials for the new buildings are commercially available on market, the main challenge comes from the existing buildings.

(3) RETs are used not only to generate electricity but also to heat and cool the indoor environment via combined heating and cooling and power solutions (tri-generation systems).

Part II
Heating Ventilation and Air Conditioning

Unit 4 Heating

Heating Pipes and User

Heating Sources

Classification of Heating System

Warm-up Activities

1. Below is a schematic diagram showing the main components of a heating system. Please match the number in the diagram to its name and description.

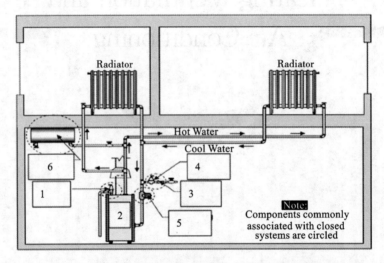

(Source: Association of Plumbing & Heating Contractors)

Terms	Number in the figure	Description
boiler		
expansion tank		
pump		
pressure reducing valve		
pressure relief valve		
back flow preventer		

Description:

a. Heat source (may be a condensing, non-condensing or a water heater).

b. Heated water will expand, and it is the place for this extra volume to go. As the system cools, the air is absorbed back into the water.

c. A device that pulls water through the system.

d. A valve that reduces potable water pressure (40-70 psi) to system pressure

(10-20 psi).

e. A spring loaded valve that opens to release heated water when internal system pressures exceed design limitations.

f. A valve that only allows water to flow one direction, typically separating the potable (drinking) water from the system (non-drinking) water.

2. Oral Exercise: Learn the following words, and tell your classmate how heat is circulated in this system.

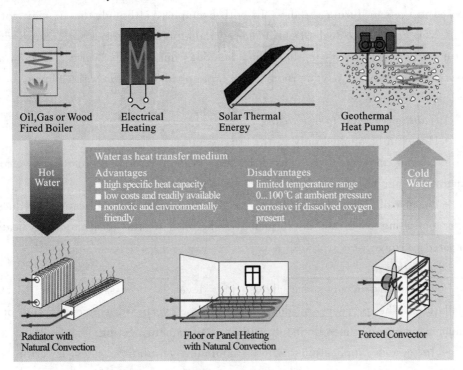

(Source: GUNT Hamburg)

Intensive Reading

Heating Equipment

There are almost endless variations and combinations of equipment that can be utilized to create the best system. Every type of heating equipment has its strength and limitations. Depending on the level of comfort and the operating costs, engineers will need to choose the style of system that integrates best with facility requirements.

Space heating engineers are usually interested in distributing heat throughout

convective heating
对流加热

the space of a building or enclosure. Most forms of space heating fall into either radiant or convective heating. Both these methods are effective in space heating and; therefore, it is important to understand the basic fundamentals.

Radiant Heating

Radiant heating systems utilize infra-red radiation to heat objects, people and surfaces [1]. Anyone who has warmed themselves by a hot wood stove or warmed their hands at a camp fire has experienced radiant heat. It is also demonstrated by standing in the sun on a winter's day; or walking near a brick wall that has been exposed to the sun during the day. In both examples, although the air may not be warm, you are able to feel the heat energy radiating from these surfaces. Radiant heat directly heats objects in the room, but does not directly warm the room air. These are appropriate, if your rooms have large open spaces or high ceilings, or are particularly draughty.

Advantages of Radiant Heating

• Heat can be applied only to the area required.

• No air movement is caused by the system itself, therefore unwelcome draughts are minimized and dust movement is reduced.

• Because of radiant heat transfer, vertical temperature stratification is reduced. Lower operating costs should be achieved because of the localization of heating compared to convection systems.

Disadvantages

• An unobstructed space above floor level is necessary for an effective installation. The presence of ductwork, pipes, overhead conveyors and other equipment may sometimes limit full utilization of radiant heating.

• In certain applications where minimum ventilation rates is critical, a combination of convection heating (or the ventilation of intake air) and radiant heating is required [2].

Convective Heating

Convective heating utilizes air circulation to transfer heat and involves two basic principles: a) cold air displaces warm air; and b) warm air rises in the presence of cold air [3]. These are either free or forced type.

• Natural convection systems rely on the buoyancy of heated air to provide circulation throughout the space. The most common examples are the steam radiator and the baseboard unit.

• Forced convection systems have a fan to force the air to circulate. Unit heaters and other fan-coil units are the common examples. These units allow the introduction of outside air and provide air filtration. To maximize efficient use of the heat energy, it is important to force the mixing and circulation of these warm air layers.

The Advantages of Convective Heating

• Convectors are used to heat up spaces more quickly than radiators.

• The convective units ensure that warm air is evenly distributed throughout the structure.

Disadvantages

• The air heaters attempt to heat the entire space including people, hardware and all of the air within the space.

• The high discharge air volumes can cause unwelcome draughts which may reduce the perceived heating effect.

• Because of their usual overhead location severe vertical temperature stratification can occur with ceiling temperatures as much as 30℃ above floor temperature [4].

• High volume air movement can also cause dust problems which could affect product quality particularly in product coating operations.

Convection heaters are appropriate if your space is insulated, well-sealed against draughts and have average ceiling heights. They should be avoided in draughty rooms, rooms with high ceilings or areas with open stairwells. Convective heating is typically the most common form of heating in majority of facilities.

Types of Heating Equipment

There are two broad classification of heating equipment.

1. **Combustion equipment**, where heat is generated by the combustion of fuel in a furnace under careful air/fuel control. The heat of combustion is recovered in some form of integral heat exchanger and is distributed via a supply air ductwork. Another form of combustion equipment is the boiler, which provides heat through a hydronic distribution system (hydronic systems are also referred to as hot water systems).

2. **Electric equipment**, where the space heaters supply heat through an electric resistance element that can convert electricity to heat with almost 100 percent efficiency. Another form is a heat pump, which extracts heat from the air, ground or water and usually delivers it through a forced air distribution system.

Distribution Systems

There are three types of distribution systems.

1. **A forced air system** circulates warmed or cooled air around the house through a network of ducts. It also provides a means of distributing ventilation air.

2. **Space heaters**, though not technically a distribution system, provide direct heat to the room in which they are located.

insulated
绝缘的
well-sealed
密封良好的

combustion equipment
燃烧设备
hydronic
循环的

3. **A hot water (hydronic) system** distributes heat through hot water pipes and radiators.

Forced Air Systems

Forced-air (convective) systems also referred to as "a central heating system" in commercial terminology utilize a series of ducts to distribute the conditioned heated or cooled air throughout the whole space [5]. The heat source is either a furnace, which burns a gas, oil or an electric heat pump. Some forced heating systems utilize hot water or steam as heating source. A blower, located in a unit called an air handler, forces the conditioned air through the ducts. Unless fresh air is piped in from outside, the system will recirculate 100% of the air. The indoor temperature is automatically controlled by a thermostat. Two important considerations are location and type. Central systems are normally controlled by a single thermostat. To achieve proper temperature control, the thermostat must be located in an area where it will sense the "average" indoor temperature. Locations exposed to localized temperature extremes (outside walls, drafts, sunlight, hot ducts or pipes, etc.) should be avoided.

Another form of forced air system is the piped system using a heating coil. Instead of a gas burner, hot water is circulated from boiler to the heating coils which heat the air. The advantages to the forced air systems are numerous. The air can be heated, cleaned, sterilized, humidified, or cooled. If return air ducts are strategically located, this will reduce heat loss by recycling the warmest air back into the system that collects at upper areas of the room [6]. Forced warm air systems have some disadvantages. Air coming from the heating registers sometimes feels cool (especially with certain heat pumps), even when it is warmer than the room temperature. There can also be short bursts of very hot air, especially with oversized units. Ductwork may transmit furnace noise and can circulate dust and odours throughout the indoor spaces. Ducts are also notoriously leaky, typically raising a space's heating costs by 20% to 30%.

Notes

[1]辐射供暖系统利用红外线辐射加热物体、人体和表面。

[2]在某些场合中,保证最低通风率是至关重要的,此时需要对流加热(或进风)和辐射加热的组合使用。

[3]对流利用空气循环来传递热量,该过程涉及两个基本原则:a)冷空气取代暖空气;b)冷空气存在时暖空气上升。

[4]由于天花板通常位于头顶,当其温度高出地板温度30℃时,可能发生严重的垂直温度分层。

[5]强制空气(对流)系统在商业术语中也称为"中央供暖系统",利用一系列管道

将调节后的加热或冷却空气分配到整个空间。

[6]如果对回风管道进行合理布局,将最热的空气再循环回房间上部收集起来,将有利于减少热损失。

Extensive Reading

Boiler

A boiler is an enclosed vessel in which water is heated and circulated, either as hot water or steam, to produce a source for either heat or power. A central heating plant may have one or more boilers that use gas, oil, or coal as fuel. The steam generated is used to heat buildings, provide hot water, and provide steam for cleaning, sterilizing, cooking, and laundering operations. Small package boilers also provide steam and hot water for small buildings.

Principles

The simple boiler is like a barrel, consisting of a cylindrical steel shell, with the ends closed by flat steel heads. It is partly filled with water and then sealed, after which a fire is started beneath it. The fire and hot gases rise around the lower outside of the shell, the heat being conducted through the steel into the water. This heats the water on the bottom of the boiler first. Hot water being lighter than cold water, rises, while the colder water in the upper part, being heavier, sinks down to replace it and is in turn heated. These are convection currents, and the process is known as circulation, which goes on continually while a boiler is in service. Circulation is good in some boilers and poor in others, depending upon the design. This is important as will be pointed out later. The water gradually reaches the temperature where steam is given off, which accumulates in the space above the water known as the steam space. As the steam accumulates, a pressure is built up which would create a very dangerous condition with the simple boiler. As pressure is exerted in every direction, the flat heads would bulge outwards because a flat surface cannot support itself. The boiler would hold very little pressure and would be useless. The first thing that has to be done with this boiler is to brace the flat heads in order to keep them from being pushed out by the pressure. This is accomplished by placing heavy steel rods, called stay rods, from head to head as shown in the next view of the simple boiler, thereby tying the heads together. The boiler can now safely carry more pressure, but it would still be an unsatisfactory boiler due to the small heating area. Improvement is made to allow more surface area of the boiler to come in contact with the hot gases of the fire by making some of the stay rods hollow and directing the hot gases through them after passing along the bottom of the shell.

The water surrounding them is heated. These hollow stay rods are called watertubes and as the fire passes through them they are called firetubes, hence the name, firetube boiler. The tubes are all located below the water level so that they are protected from the heat.

Boiler Rating

Boilers are rated in kilowatts, where 1 watt equates to 1 Joule of energy per second, i.e. 1 W = 1 J/s. Many manufacturers still use the imperial measure of British thermal units per hour for their boilers. For comparison purposes 1 kW equates to 3412 Btu/h.

Rating can be expressed in terms of gross or net heat input into the appliance. Values can be calculated by multiplying the fuel flow rate (m^3/s) by its calorific value (kJ/m^3 or kJ/kg). Input may be gross if the latent heat due to condensation of water is included in the heat transfer from the fuel. Where both values are provided in the appliance manufacturer's information, an approximate figure for boiler operating efficiency can be obtained, e.g. if a gas boiler has gross and net input values of 30 kW and 24 kW respectively, the efficiency is $24/30 \times 100\% = 80\%$.

Oil and solid fuel appliances are normally rated by the maximum declared energy output (kW), whereas gas appliances are rated by net heat input rate (kW net).

Calculation of boiler power:

$$kW = \frac{kg \text{ of water} \times S.h.c. \times Temp. \text{ rise}}{Time \text{ in seconds}} \quad (1)$$

where:

1 litre of water weighs 1 kg;

S.h.c. = specific heat capacity of water, $4.2 \text{ kJ/(kg} \cdot \text{K)}$;

Temp. rise = rise in temperature that the boiler will need to increase the existing mixed water temperature (say 30℃) to the required storage temperature (say 60℃);

Time in seconds = time the boiler takes to achieve the temperature rise. 1 to 2 hours is typical, use 1.5 hours in this example.

From the example on the previous page, storage capacity is 1500 litres, i.e. 1500 kg of water. Therefore:

$$\text{Boiler power} = \frac{1500 \times 4.2 \times (60-30)}{1.5 \times 3600} = 35 \text{ (kW net)}$$

Given the boiler has an efficiency of 80%, it will be gross input rated 43.75 kW.

▶▶▶ Post-Reading Exercises

1. Fill in the blanks with the words or expressions given below, change the

form where necessary.

The simple boiler _____ of a cylindrical steel shell _____ the ends closed by flat steel heads. Boiler will be heated _____ it. This heats the water on the bottom of the boiler first. Hot water rises, while the colder water in the upper part _____ down. These combine the effect of conduction and convection.

The water gradually reaches the temperature where steam is _____. As the steam accumulates, pressure is built up, and the flat heads would bulge _____. The boiler would hold very little pressure and would be useless.

The first thing that has to be done with this boiler is to brace the flat heads in order to keep them from being pushed out by the pressure. This is _____ by placing stay rods. The boiler can now safely carry more pressure, but it would still be an unsatisfactory boiler due to the _____ heating area. More surface area of the boiler shall be made to contact _____ the hot gases of the fire by making some of the stay rods hollow and _____ the hot gases through them after passing along the bottom of the shell.

| outwards | with | beneath | direct | give off |
| limit | accomplish | consist | whose | sink |

2. Answer the following questions.
(1) What is the difference of watertube boiler and firetube boiler?

(2) Why does not the boiler have flat heads?

3. Translation.
(1) The water gradually reaches the temperature where steam is given off, which accumulates in the space above the water known as the steam space.

(2) Rating can be expressed in terms of gross or net heat input into the appliance. Values can be calculated by multiplying the fuel flow rate (m^3/s) by its calorific value (kJ/m^3 or kJ/kg).

Combined Heat and Power

There is an increasing interest in Cogeneration or Combined Heat and Power

(CHP) to reduce the global warming effects of the use of fuels for heating, as this technology provides a way to use heat to heat buildings, that otherwise would be rejected to the environment as part of the conversion of fuel to power and electricity.

All power plants, transport vehicles and utility-scale electrical power stations, transform less than half the energy content of their input fuel into electricity. The rest of the fuel's energy they "reject" or "waste" as heat to the environment, typically to river- or sea-cooling water bodies or through cooling towers and exhaust stacks.

The amount of fuel usefully converted to electricity is defined by the percentage of electricity or power produced per unit of energy in the input fuel. This efficiency depends on the thermodynamic cycle used for the conversion. One of the most efficient thermodynamic cycles for conversion is the combined cycle gas turbine (CCGT) power station which combines the use of a gas turbine with a steam turbine. Such plants convert about 60% of the fuel energy to electrical power. In contrast the steam turbine cycle for nuclear fuel, biomass or coal has efficiencies in the range of 34-40%. Cycles using steam turbines are most effective when they reject the waste heat at as low a temperature as possible to the environment typically at around 30℃, a temperature too low for practical large-scale heating purposes, but useful for horticulture or fish farming and local underfloor heating.

The reject heat in a CHP plant satisfies a heat demand, such as heating a factory process or heating buildings, where this would otherwise require energy from typically another fuel, burnt in a heat-only boiler.

CHP technology covers a very broad range of both technologies and sizes, from 1 kW electrical output unit to 400 MW. Technologies can include steam turbines, gas turbines, engines, combined cycles, micro-turbines, fuel cells and others.

Total losses from the European energy system in 2008 were 37% of primary energy input and are largely from the electricity generating sector. Analysis shows that CHP can provide large primary energy savings in comparison to the conventional production of electricity and heat in separate plants, albeit at an extra capital cost typically of 10% up to 25% of the electricity-only station. The extra capital cost is usually repaid in a commercially realistic time frame particularly for industry and where CO_2 heat, is usually provided in the form of hot exhaust gases, steam or hot water, and sometimes thermal oil. Fossil, nuclear, waste and renewable fuels can be used to power CHP plants.

CHP is about power, so where the power directly drives mechanical equipment such as a heat pump, a water pump or fan, this is also a more efficient form of CHP, as it does not incur the associated losses of converting power to electricity and then back to power again. Use of the reject heat can also replace the burning of fuel or use of other heat sources in absorption cooling cycles. Some cities, as a result, have district cooling using the process and other methods to deliver cooling.

Operating Principles

CHP can be applied in a range of power generating technologies. In each case the power generating technology is available as an electricity-only generator. For the largest units, heating cities, these are designed so that they can just produce electricity at times when no heat is required, rejecting heat to the environment at 30℃ or alternatively, producing electricity and low-carbon heat at temperature 80-95 ℃ suitable for city heating when both heat and electricity are required.

Typical CHP technologies include: steam turbines, gas turbines, combined cycle gas turbines CCGTs (a combination of the first two plant types) and gas engines (similar to a car engine). Other more niche technologies include: organic Rankine cycle (ORC) turbines (similar to steam turbines but using an organic fluid rather than steam) which are suitable for small (1-3 MW) biomass combustion plants, diesel engines, micro turbines (i.e. gas turbines below about 50 kW) and stirling engines. Work proceeds on commercialising fuel cell CHP.

In power cycles, a working fluid such as steam, air, hydrogen or an organic compound vapour, is subjected to a thermodynamic cycle, i.e. a gas is compressed, heated and then expanded when work is done and power generated followed by heat rejection to cool the working fluid.

In CHP, the steam cycle may be modified so that the heat is rejected at a sufficiently high temperature to be used for a separate heating purpose and some of the plant details may be changed. In gas engines or gas turbines, waste heat is readily available at high temperature and use of this heat has no effect on power output or efficiency. In both cases, extra heat exchangers are fitted to recover the various waste-heat streams and to transmit them to the heating medium.

The steam cycle plant can be operated as if it were a normal electricity-only power station, in which case all the steam from the turbine is cooled in a condenser and turned from steam to water giving up its latent heat at around 30 ℃. This is referred to as fully condensing mode maximising the power from the steam. When using it as CHP, there is an option to extract some of the steam at a higher temperature and pressure and feed it to a district heating condenser containing city heating water. When this happens, the electrical output of the power station will drop as a small amount of the energy in the steam is lost by condensing it at a higher temperature, but the fuel consumption remains constant.

The fuel used to enable the waste heat to be at a useful temperature is measured by the Turbine Z factor. Typically, a loss of 1 unit of electricity output will result in between 5 and 10 (depending on the power plant) units of heat becoming available at a usable temperature. The higher the temperature at which the heat is required, generally the lower the Z factor will be. This is identical to the way an electric heat pump reduces its heat delivery per unit of power used, with the difference between its source heat temperature and its delivery heat temperature.

Comparison of Performance with Electric Heat Pumps and CHP

The ratio of power loss to heat gained compares very favourably with an electric heat pump, which typically will use 1 unit of electricity to provide 3 units of heat, i.e. has a coefficient of performance (COP) of 3. Thus, CHP may be considered as a virtual steam-cycle heat pump, i.e. a loss of electricity to make heat available. According to theoretical studies, if the lowest possible DH temperature is chosen and multi-stage steam extraction is used, then COP of 18 can be achieved.

▶▶▶ Post-Reading Exercises

1. Rephrase the following passage without changing its meaning.

In CHP, the steam cycle may be modified so that the heat is rejected at a sufficiently high temperature to be used for a separate heating purpose and some of the plant details may be changed. In gas engines or gas turbines, waste heat is readily available at high temperature and use of this heat has no effect on power output or efficiency. In both cases, extra heat exchangers are fitted to recover the various waste-heat steams and to transmit them to the heating medium.

2. Choose the best answer for each of the following.

(1) The energy-to-power efficiency of steam turbine cycle for nuclear fuel is _____.
   A. 60%
   B. 34-40%
   C. higher than 50%

(2) Compared to the electricity-only station, CHP can provide large primary energy savings at an extra capital cost of _____.
   A. 10-25%
   B. 5-10%

C. 25-30%
(3) Steam turbines are most effective when they reject the waste heat at as low a temperature as possible to the environment typically at around 30℃, which is not suitable for _____.
A. fish farming
B. horticulture
C. regional heating
(4) In power cycles, a working fluid is _____.
A. compressed-expanded-heated
B. compressed-heated-expanded
C. heated-expanded-compressed
(5) Combined cycle gas turbine power stations convert about _____ of the fuel energy to electrical power.
A. 60%
B. 50%
C. 40%

Unit 5　Ventilation

Mechanical Ventilation

Warm-up Activities

1. Learn the following components of a typical ventilation system and their functions.

  1. Air Intake
  2. Insulation Valve
  3. Filter
  4. By-pass
  5. Air Treatment(optional)
  6. Starting Valve
  7. Fan
  8. Air Duct
  9. Outlet/Diffuser
  10. Regulating Valve
  11. Duct

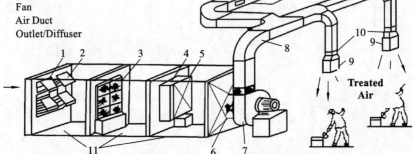

2. Oral Exercise: Prepare a short presentation about "Necessity of Ventilation".

Intensive Reading

General ventilation

contaminant
污染物
louver
百叶窗
staleness
腐旧的

General ventilation controls heat, odours, and contaminants. It may be provided by natural draft, by a combination of general supply and exhaust air fan and duct systems, by exhaust fans only (with replacement air through inlet louvers and doors), or by supply fans only (exhaust through relief louvers and doors) [1].

It is important to provide at least a minimum amount of fresh air indoors, both for comfort and for health. Odours and a sense of staleness can be uncomfortable, and build-ups of pollutants can be produced within buildings. These pollutants are easily removed with air changes through rooms.

Winter heat loss (and summer heat gain in closed, cooled buildings) occurs when fresh outdoor air enters a building to replace stale indoor air. This heat exchange must be calculated when sizing heating or cooling equipment or when estimating energy use per season.

Air exchange increases a building's thermal load in three ways. First, the incoming air must be heated or cooled from the outdoor air temperature to the indoor air temperature. Second, air exchange increases a building's moisture content, which means humid outdoor must be dehumidified. Third, air exchange can increase a building's thermal load by decreasing the performance of the envelope insulation. Air flowing around and through the insulation can increase heat transfer rates above design rate. Air flow within the insulation system can also decrease the system's performance due to moisture condensing in and on the insulation.

moisture condensing
水汽凝结

Ventilation Requirements

Requirements for an acceptable amount of fresh air supply in buildings will vary depending on the nature of occupation and activity. As a guide, 10 L/s of outdoor air supply per person can be applied for a non-smoking environment, to an extract air rate of 36 L/s per person in a room dedicated specifically for smokers, equates to 36 to 130 m^3/h per person [2].

Ventilation Types

Ventilation moves outdoor air into a building or a room, and distributes the air within the building or room. The general purpose of ventilation in buildings is to provide healthy air for breathing by both diluting the pollutants originating in the building and removing the pollutants from it [3].

dilute
稀释

Building ventilation has three basic elements.

• ventilation rate — the amount of outdoor air that is provided into the space, and the quality of the outdoor air;

• airflow direction — the overall airflow direction in a building, which should be from clean zones to dirty zones; and

• air distribution or airflow pattern — the external air should be delivered to each part of the space in an efficient manner and the airborne pollutants generated in each part of the space should also be removed in an efficient manner.

airborne pollutant
空气污染物

There are three methods that may be used to ventilate a building: natural, mechanical and hybrid (mixed-mode) ventilation.

• Natural ventilation is driven by pressure differences between one part of a building and another, or pressure differences between the inside and outside.

• Mechanical (or forced) ventilation is driven by fans or other mechanical plant.

Natural Ventilation

Natural ventilation is generally preferable to mechanical ventilation as it will typically have lower capital, operational and maintenance costs. However, there are a range of circumstances in which natural ventilation may not be possible, for instance, the building is too deep to ventilate from the perimeter or windows cannot be opened, etc. Some of these issues can be avoided or mitigated by careful design, and mixed mode or assisted ventilation might be possible, where natural ventilation is supplemented by mechanical systems.

Types of Mechanical Ventilation

Where mechanical ventilation is necessary it can be:

• A circulation system such as a ceiling fan, which creates internal air movement, but does not introduce fresh air.

• A pressure system, in which fresh outside air is blown into the building by inlet fans, creating a higher internal pressure than the outside air.

• A vacuum system, in which stale internal air is extracted from the building by an exhaust fan, creating lower pressure inside the building than the outside air.

• A balanced system that uses both inlet and extract fans, maintaining the internal air pressure at a similar level to the outside air and so reducing air infiltration and draughts.

• A local exhaust system that extracts local sources of heat or contaminants at their source, such as cooker hoods, fume cupboards and so on.

In commercial developments, mechanical ventilation is typically driven by air handling units(AHU) connected to ductwork within the building that supplies air to and extracts air from interior spaces. Typically, AHU comprise an insulated box that forms the housing for filter racks or chambers, a fan (or blower), and sometimes heating elements, cooling elements, sound attenuators and dampers. In some situations, such as in swimming pools, air handling units might include dehumidification.

Calculating Air Change Rates

The design of mechanical ventilation systems is generally a specialist task, undertaken by a building services engineer. Air is continuously exchanged between buildings and their surroundings as a result of mechanical and passive ventilation and infiltration through the building envelope. The rate at which air is exchanged is an important property for the purposes of ventilation design and heat loss calculations and is expressed in "air changes per hour" (ach).

If a building has an air change rate of 1 ach, this equates to all of the air within the internal volume of the building being replaced over a 1 hour period.

A number of techniques are available for calculating the air change rate of a building. The choice of method depends on the accuracy required. The most straightforward method relies on the use of a simple mathematical equation, while the most complex methods use computational analysis and consider many different variables (such as computational fluid dynamics). The basic method calculates air change rates using the following equation:

$$n = \frac{3600q}{V} \tag{1}$$

where:

n = Air changes per hour (ach);
q = Fresh air flow rate (m³/s);
V = Volume of the room (m³).

Whilst there are standards and rules of thumb that can be used to determine air flow rates for straightforward situations, when mechanical ventilation is combined with heating, cooling, humidity control and the interaction with natural ventilation, thermal mass and solar gain, the situation can quickly become very complicated [4].

This, along with additional considerations, such as the noise generated by fans, and the impact of ductwork on acoustic separation means it is vital that building services are considered at the outset of the building design process [5].

acoustic
声学的

Notes

[1]一般通风可通过自然通风、送排风风机、管道系统的组合使用，或仅通过排风机（通过入口百叶窗和门更换空气）或送风机（通过安全百叶窗和门排气）来实现。

[2]作为指导，在禁烟环境下，提供每人 10 L/s 的室外空气，而在吸烟室内，保证每人 36 L/s 的抽气速率，相当于每人 36～130 m³/h。

[3]建筑物通风的一般目的是通过稀释建筑物内的污染物和去除建筑物内的污染物，提供健康的空气。

[4]虽然有一些标准和经验法则可用于确定简单情况下的空气流量，但当机械通风与加热、冷却、湿度控制相结合，与自然通风、热质量和太阳能获得量相互作用时，情况可能很快变得非常复杂。

[5]这一点，加上额外的考虑因素，如风机产生的噪声，以及管道系统对声学分离的影响，意味着在建筑设计过程开始时就考虑建筑设备是至关重要的。

Extensive Reading

Displacement Ventilation

To avoid confusion, it is necessary to understand what makes an HVAC system a displacement ventilation (DV) system. The *ASHRAE Handbook of Fundamentals* (HOF) classifies room air diffusion systems as mixing, displacement, unidirectional, and underfloor. The HOF describes mixing systems as normally discharging conditioned air from outlets near either the ceiling or floor at velocities much greater than those acceptable in the occupied zone to achieve a level of air mixing that creates relatively uniform air velocity, temperature, humidity, and air quality conditions in the occupied zone. Most non-industrial buildings in the U. S. use mixing systems.

In contrast, the HOF describes DV systems as supplying slightly cool air at low velocities directly to the occupied zone from outlets at or near the floor level. The supply air spreads over the floor, forms thermal plumes upon encountering heat sources, and is exhausted by system returns located at or near the ceiling. DV systems purposefully minimize mixing and, in fact, are designed to establish a stable thermal stratification level above the occupied zone such that no mixing occurs between the upper and lower zones. Although not discussed in the HOF, DV systems, as typically applied in Europe, do not employ recirculation of room exhaust air.

Underfloor air distribution (UFAD) systems supply air through a raised floor to local areas typically near building occupants and return air at or near the ceiling. Since both UFAD and DV systems utilize a similar low supply paired with high return pattern, the two are often confused. However, UFAD systems differ from DV systems because they typically supply air at higher velocities such that their higher supply volumes are able to meet larger cooling loads. Depending on the design details of the system and space, UFAD systems may operate as DV systems, however, UFAD systems are outside the scope of this report. McDonnell (2003) recently described UFAD and DV systems with emphasis on key differences. ASHRAE has recently published a detailed guide on UFAD system design that covers topics such as thermal comfort, ventilation effectiveness, energy use, and practical guidance on equipment selection, design, construction, commissioning, operation and maintenance issues.

The key performance issue for successful DV application is unidirectional flow and the establishment of a stable thermal stratification layer within the zone. As such, many researchers have studied the room airflow, temperature, and contaminant concentration patterns resulting from DV systems through measurements, computational fluid dynamics (CFD), or other modeling approaches.

Chapter 3 of the ASHRAE guidebook summarizes the basic features of DV performance in nonindustrial applications in terms of room airflow patterns, temperature distribution, contaminant distribution and thermal comfort as informed by the European research and 20 years of practical experience. This chapter describes desirable DV system operation as stratification leading to two stable zones — a cooler, cleaner zone ending at a boundary somewhere above the occupant breathing zone and a warmer, more contaminated zone above the boundary. Plumes from occupants and other heat sources effectively transport both heat and contaminants from the lower zone to the upper zone.

Many factors affecting the establishment of such a stratified space include location and strength of both heat and contaminant sources, supply air temperature and airflow rate, room geometry, warm or cool walls and ceilings, and existence of infiltration air. Chapter 4 of the guidebook also discusses the critical topic of draft problems when supplying cool air at or near the floor of occupied spaces. The guidance emphasizes careful diffuser selection to achieve the desired performance.

The ASHRAE guidebook also describes new research that addressed the DV system performance issues in the specific context of expected higher cooling loads in U.S. buildings due to both climate and internal gains relative to Northern European buildings. Chen and Glicksman (1999) describe this research in detail including experimental measurements of DV in a test room, validation of a CFD model, CFD predictions of DV performance (in a small office, a large office with internal partitions, a classroom, and a workshop), development of simple models of temperature difference and ventilation effectiveness, and energy and cost analysis. They report good agreement between measured and predicted air velocities and temperatures but greater discrepancies for contaminant concentrations. Simplified models were developed for the vertical temperature distribution and ventilation effectiveness, however, the ventilation effectiveness model only considered occupants as contaminant sources (i.e., associated with heat sources). The ventilation effectiveness model also showed that ventilation effectiveness decreased at lower ventilation rates, with the decrease being very pronounced at sufficiently low ventilation rates. Additionally, the ventilation effectiveness depended on the fraction of the heat load located in the occupied zone of the room. Based on the CFD studies, the authors conclude that properly designed DV systems can maintain thermal comfort while achieving better IAQ than mixing ventilation.

▶▶▶ Post-Reading Exercises

1. Scan the text and fill in the blanks.

(1) According to HOF, DV systems supply slightly cool air at low velocities to _____ from outlets at or near the floor level. The air is exhausted by returns located _____.

(2) DV system creates 2 zones: _____ and _____.

(3) Ventilation effectiveness depends on _____.

(4) In DV systems, mixing is _____.

(5) A stratified space is affected by _____, _____, _____, _____ and _____.

2. Translation.

(1) The HOF describes mixing systems as normally discharging conditioned air from outlets near either the ceiling or floor at velocities much greater than those acceptable in the occupied zone to achieve a level of air mixing that creates relatively uniform air velocity, temperature, humidity, and air quality conditions in the occupied zone.

(2) Chen and Glicksman (1999) describe this research in detail including experimental measurements of DV in a test room, validation of a CFD model, CFD predictions of DV performance (in a small office, a large office with internal partitions, a classroom, and a workshop), development of simple models of temperature difference and ventilation effectiveness, and energy and cost analysis.

Smoke Control

One of the most hazardous situations that can be faced in a building is smoke. While fires themselves are often damaging, it is smoke that can cause the most injuries.

In order to protect a building's occupants, as well as furnishings and equipment that may be damaged by smoke, a smoke control system is needed. A smoke control system controls the flow of smoke in a building in the event of a fire. It keeps smoke from spreading throughout the building and gives the building's occupants a clear evacuation route, as well as preventing further damage to the building's interior.

Building Pressurization

The primary means of controlling smoke movement is creating air pressure differences between smoke control zones. The basic concept of building pressurization is to establish a higher pressure in adjacent spaces than in the smoke zone. In this way, air moves into the smoke zone from the adjacent areas and smoke is prevented from dispersing throughout the building.

Dedicated and Non-Dedicated Systems

Smoke control systems are either dedicated or non-dedicated. A dedicated smoke control system is installed in a building for the sole purpose of controlling smoke. It is a separate system of air moving and distribution equipment that does not function under normal building operating conditions. Dedicated systems are used for special areas, such as elevator shafts and stair towers that require special smoke control techniques. Non-dedicated smoke control systems are systems that share components with some other systems such as the building automation (HVAC) system. When activated, the system changes its mode of operation to achieve the smoke control objectives.

Fire Control Systems

The goal of a fire control system is to contain and extinguish the fire as fast as possible. Fire control systems halt the fire, but not the smoke, and are often triggered automatically by the heat of the fire. These systems rely on a water supply, such as sprinklers, whereas smoke control systems usually rely on electricity to run fans and dampers.

The smoke control system is usually separated from the fire control system because they have different goals. However, the smoke control system should be designed to work with the fire control system and not interfere with its operation. For example, if the building has a sprinkler system, then the smoke control system does not need to control a large quantity of smoke because the size of any fire should be smaller.

Moreover, if a smoke control system is working with a gas-based fire extinguisher, certain actions must be taken. If the smoke control system tried to vent a room with a gas-based fire extinguishing system, it would probably vent the smoke along with the fire suppressing gas. Removing the gas lets the fire continue burning. Therefore, gas-based fire extinguishers and smoke control systems should not be active at the same time in the same area.

Smoke control systems receive the location of the fire from the fire panel. The fire panel uses a combination of smoke and heat sensors to determine where the fire is located. In the event that signals are received from more than one smoke zone, the smoke control system should continue automatic operation in the mode determined by the first signal received.

Shaft Protection

There are two types of shaft protection systems:
- Stairwell pressurization systems.
- Elevator smoke control.

Stairwell Pressurization Systems

Stair towers are stairwells with a ventilation system and are isolated from the main building. Stair towers are the most common type of dedicated smoke control system. The only connection between the building and the stair tower is the fire-rated doors on each floor. Because the building's occupants should use the stair tower to leave during an evacuation, keeping the stair tower smoke free is vital. The following image (Fig. 5-1) depicts the various parts of a stair tower system.

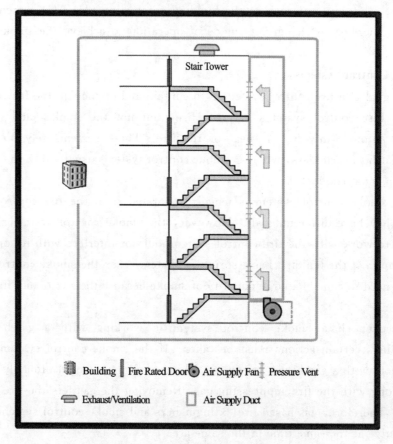

Figure 5-1 Stair tower system components

The goal of pressurized stairwells is to maintain a tenable environment within exit stairwells for the time necessary to allow occupants to exit the building. A tenable environment is defined as an environment in which the products of combustion, including toxic gases, particulates, and heat, are limited or otherwise restricted to maintain the impact on occupants to a level that is not life threatening. A secondary objective of stairwell pressurization is to provide a staging area for firefighters. This is achieved when stair shafts are mechanically pressurized, with respect to the fire area, with outdoor air to keep smoke from contaminating them during a fire. It is important to pressurize a stair tower enough to keep smoke out. However, if the pressure in the stair tower is too great, opening the door leading into the stair tower can be difficult.

Elevator Smoke Control

Elevator shafts present a particular problem with regard to smoke control. The elevator shafts form perfect chimneys to draw smoke into the upper levels of a building. Since elevators usually have openings on each floor, and the seals on elevator doors are often poor, the elevator shaft can become a mechanism to spread smoke throughout a building.

In order to have a usable elevator during a smoke emergency, the elevator shafts have to be pressurized in the same way as a stair tower is pressurized. However, pressurizing the elevator shaft presents a number of problems. You can fit the elevator doors with improved seals and rubber sweeps. However, these steps do not completely eliminate air leakage. Moreover, most elevator shafts are not designed to be pressurized. There are also localized pressure differences that the cars create as they travel up and down the shafts. Shafts are often constructed of porous material that cannot contain the air pressure. Shafts are not designed for inspection after the elevators are installed, so finding and repairing cracks that would let smoke infiltrate or pressure escape is difficult.

While several methods for correcting the problems of air pressurization in elevators have been proposed and investigated, there are no firm recommendations regarding elevator smoke control. Refer to the NFPA 92 and UL 864 standards for additional information. Remembering the local AHJ and project specifications may require control beyond that specified in the above standards.

Floor Protection

As discussed previously, pressurized stairwells are intended to control smoke to the extent that they inhibit smoke infiltration into the stairwell. However, in a building with just a pressurized stairwell, smoke can flow through cracks in floors and partitions and through other shafts to threaten life or damage property at locations remote from the fire. The concept of zoned smoke control is intended to limit this type of smoke movement from within a building.

With zoned smoke control, smoke movement is inhibited by dividing the building into smoke control zones, with each zone separated from the others by smoke barriers. These smoke barriers can include partitions, floors, or doors that you can close. When a fire occurs in one of these smoke control zones, it is called a smoke zone. In the event of a fire, pressure differences and airflows produced by mechanical fans and operating dampers can limit the smoke to the zone in which the fire originated. When a fire/smoke condition occurs, all of the non-smoke zones that are contiguous to the smoke zone are positively pressurized and the smoke zone is negatively pressurized. Optionally, all of the remaining smoke control zones in the building may also be positively pressurized. With the smoke contained to the smoke zone, it can then be exhausted. Typically, the fire/smoke signals from a protective signalling system are used to activate the zoned smoke control sequence.

▶▶▶ Post-Reading Exercises

1. Scan the text and fill in the blanks.

(1) The way of controlling smoke movement is _____.

(2) The building and the stair tower are connected only by _____.

(3) Smoke zones are separated from each other by _____.

2. True or False.

(1) A smoke control system can keep the building from fire.   (   )

(2) Non-dedicated smoke control system's air moving and distribution equipment does not function under normal building operating conditions.   (   )

(3) Dedicated control system shares components with some other systems.
                                                                (   )

(4) The smoke control system is totally separated from the fire control system.
                                                                (   )

3. Short answer.

(1) Why cannot the gas-based fire extinguisher and the smoke control system simultaneously be active?

(2) What is a "tenable environment"?

(3) Describe the pressure difference between the smoke zones when a fire occurs in one of them.

Unit 6　Air Conditioning

Warm-up Activities

1. Learn the following words and find out what kind of AC system it represents.

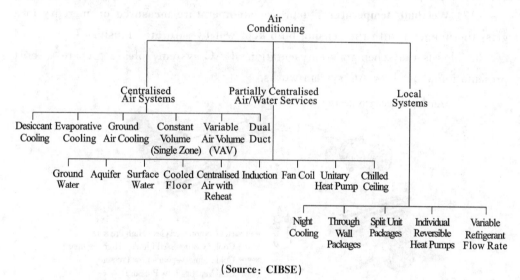

(Source: CIBSE)

2. The following is a glossary of terminology used in air conditioning design. Please translate the following sentences into Chinese.

(1) **Dew point**: temperature at which the air is saturated (100% RH) and further cooling manifests in condensation from water in the air.

(2) **Dry bulb temperature**: temperature shown by a dry sensing element such as mercury in a glass tube thermometer.

(3) **Enthalpy**: total heat energy, i.e. sensible heat + latent heat.

(4) **Entropy**: measure of total heat energy in a refrigerant for every degree of temperature.

(5) **Latent heat**: heat energy added or removed as a substance changes state, whilst temperature remains constant, e.g. water changing to steam at 100 ℃ and atmospheric pressure (W).

(6) **Moisture content**: amount of moisture present in a unit mass of air (kg/kg dry air).

(7) **Percentage saturation**: ratio of the amount of moisture in the air compared with the moisture content of saturated air at the same dry bulb temperature. Almost the same as RH and often used in place of it.

(8) **Relative humidity (RH)**: ratio of water contained in air at a given dry bulb temperature, as a percentage of the maximum amount of water that could be held in air at that temperature.

(9) **Saturated air**: air at 100% RH.

(10) **Sensible heat**: heat energy which causes the temperature of a substance to change without changing its state (W).

(11) **Specific volume**: quantity of air per unit mass (m^3/kg).

(12) **Wet bulb temperature**: depressed temperature measured on mercury in a glass thermometer with the sensing bulb kept wet by saturated muslin (℃ wb).

3. This is the schematic representation of AC system, please prepare a short presentation about how AC system works.

How It Works

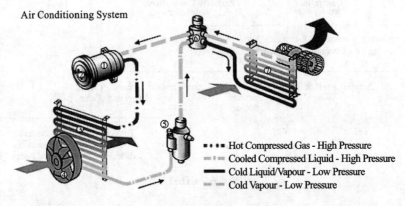

(Source: VDS performance)

Intensive Reading

Air Conditioning and Refrigeration

refrigeration
制冷
refrigerant
制冷剂
heating medium
加热介质

A complete air-conditioning system includes a means of refrigeration, one or more heat transfer units, air filters, a means of air distribution, an arrangement for piping the refrigerant and heating medium, and controls to regulate the proper capacity of these components [1]. In addition, the application and design requirements that an air-conditioning system must meet make it necessary to arrange some of these components to condition the air in a certain sequence. For example, an installation that requires reheating of the conditioned air must be arranged with the reheating coil on the downstream side of the dehumidifying coil; otherwise, it is impossible to reheat the cooled and dehumidified air. There has been a tendency by many designers to classify an air-conditioning system by referring to one of its components. For example, the air-conditioning system that includes a dual duct arrangement to distribute the conditioned air is referred to as a dual duct system. This classification makes no reference to the type of refrigeration, the piping arrangement, or the type of controls.

Basic Components of an HVAC System

An HVAC System may include the following basic components or units (Fig. 6-1).
- HVAC water chillers and heaters.
- Hot water generator (if chiller does produce chilled water only).
- Chilled water pumps.
- Cooling water pumps.
- Electrical power supply control or motor control center (MCC).
- Cooling towers.
- Piping for chilled water and cooling water or condenser side water.
- Valves for chilled water and cooling water sides.
- Air handling units (AHUs), heating coils and cooling coils.
- Ducts in ventilation system (supply ducts and return ducts).
- Fan Coil Units (FCUs) and thermostats.
- HVAC diffusers and grills.
- HVAC controls (instrumentation & control components) installed at various locations.
- HVAC software for building HVAC control or building management system (BMS).

An assembly of all above components forms an HVAC system.

Classification

The major classification of HVAC systems is central system and decentralized or local system. Types of a system depend on addressing the primary equipment location to be centralized as conditioning entire building as a whole unit or decentralized as separately conditioning a specific zone as part of a building [2]. Therefore, the air and water distribution system should be designed based on system classification and the location of primary equipment. The criteria as mentioned above should also be applied in selecting between two systems. Table 6-1 shows the comparison of central and local systems according to the selection criteria.

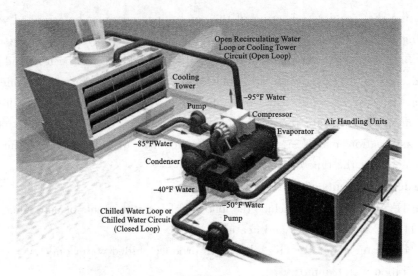

Figure 6-1 Basic components of an HVAC system

Table 6-1 Comparison of Central and Local HVAC Systems

Criteria	Central system	Decentralized system
Redundancy	Standby equipment is accommodated for troubleshooting and maintenance	No backup or standby equipment
First cost	High capital cost	Affordable capital cost
Operating cost	More significant energy efficient primary equipment	Less energy efficient primary equipment
Maintenance cost	Accessible to the equipment room for maintenance and saving equipment in excellent condition, which saves maintenance cost	Accessible to equipment to be located in the basement or the living space. However, it is difficult for roof location due to bad weather
Reliability	Central system equipment can be an attractive benefit when considering its long service life	Reliable equipment, although the estimated equipment service life may be less
Flexibility	Selecting standby equipment to provide an alternative source of HVAC or backup	Placed in numerous locations to be more flexible

standby equipment
备用设备

capital cost
资本成本

Classification of AC and VRF System

Central HVAC Systems

A central HVAC system may serve one or more thermal zones, and its major equipment is located outside of the served zone(s) in a suitable central location whether inside, on top, or adjacent to the building [3]. Central systems must condition zones with their equivalent thermal load. Central HVAC systems will have several control points such as thermostats for each zone. The medium is used in the control system to provide the thermal energy sub-classifies the central HVAC system. The thermal energy transfer medium can be air or water or both, which can represent as all-air systems, air-water systems, all-water systems. Also, central systems include water-source heat pumps and heating and cooling panels. Central HVAC system has combined devices in an air handling unit which contains supply and return air fans, humidifier, reheat coil, cooling coil, preheat coil, mixing box, filter, and outdoor air.

All-Air System
all-air system
全空气系统
all-water system
全水系统

All-Air Systems

The thermal energy transfer medium through the building delivery systems is air. All-air systems can be sub-classified based on the zone as single zone and multizone, airflow rate for each zone as constant air volume and variable air volume, terminal reheat, and dual duct.

Some spaces require different airflow of supply air due to the changes in thermal loads. Therefore, a variable-air-volume (VAV) all-air system is the suitable solution for achieving thermal comfort. The VAV system consists of a central air handling unit which provides supply air to the VAV terminal control box that located in each zone to adjust the supply air volume [4]. The temperature of supply air of each zone is controlled by manipulating the supply air flow rate. The main disadvantage is that the controlled airflow rate can negatively impact other adjacent zones with different or similar airflow rate and temperature. Also, part-load conditions in buildings may require low air-flow rate which reduces the fan power resulting in energy savings. It may also reduce the ventilation flow rate, which can be problematic to the HVAC system and affect the indoor air quality of the building.

variable air volume
变风量

terminal control box
终端控制箱

All-Water Systems

In an all-water system, heated and cooled water is distributed from a central system to conditioned spaces. This type of system is relatively small compared to other types because the use of pipes as distribution containers and the water has higher heat capacity and density than air, which requires the lower volume to transfer heat. All-water heating-only systems include several delivery devices such as floor radiators, baseboard radiators, unit thermostat which controls the water flow to the fan-coil units. In addition, occupants can adjust fan coil units by adjusting supply air louvers to achieve the desired temperature. The main disadvantage of fan-coils is ventilation air and only can be solved if the fan-coil units (Fig. 6-2) are

All-Water System

connected to outdoor air. Another disadvantage is the noise level, especially in critical places.

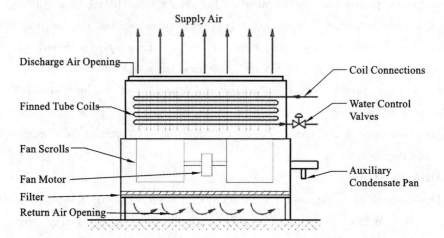

Figure 6-2 All-water system: fan-coil units

Air-Water Systems

Air-water systems are introduced as a hybrid system to combine both advantages of all-air and all-water systems. The volume of the combined is reduced, and the outdoor ventilation is produced to properly condition the desired zone. The water medium is responsible for carrying the thermal load in a building by 80-90% through heating and cooling water, while air medium conditions the remainder. There are two main types: fan-coil units and induction units.

Water-Source Heat Pumps

Water-source heat pumps are used to provide considerable energy savings for large building under the extreme cold weather. A building of various zones can be conditioned by several individual heat pumps since each heat pump can be controlled according to the zone control. A centralized water circulation loop can be used as a heat source and heat sink for heat pumps. Therefore, heat pumps can act as the primary source of heating and cooling. The main disadvantage is the lack of air ventilation similar to the all-water systems as in fan-coil units. For a heating process, the boiler or solar collectors will be used to supply heat to the water circulation, while a cooling tower is used to reject heat collected from the heat pumps to the atmosphere. This system does not use chillers or any refrigeration systems. If a building requires a heating process for zones and cooling process for other zones at the same time, the heat pump will redistribute heat from one part to another with no need for a boiler or cooling tower operation [5].

Heating and Cooling Panels

Heating and cooling panels are placed on floors or walls or ceilings which can be a source of heating and cooling. It also can be called as radiant panels. This type of system can be constructed as tubes or pipes impeded inside the surface where the cooling or heating media is circulated into the tubes to cool or heat the surface. The tubes are contacted to the adjacent large surface area to achieve the desired surface temperature for cooling and heating process. The heat transfer process is mainly by the radiation mode between the occupants and the radiant panels, and the natural convection mode between the air and panels [6]. Temperature restriction is recommended for radiant floor panels, a range of 66-84°F, to achieve thermal comfort for occupants (ASHRAE Standard 55). Radiant ceiling or wall panels can be used for cooling and heating process. The surface temperature should be higher than the air dew point temperature to avoid condensation on the surface during the cooling process. Also, the maximum surface temperature is 140°F for ceiling levels at 10 ft. and 180°F for ceiling levels at 18 ft. This temperature is recommended to avoid too much heating above occupants' heads.

radiant panel
辐射板

Notes

[1] 一个完整的空调系统包括制冷装置、一个或多个传热装置、空气过滤器、空气输配装置、冷热媒管道以及适当的控制装置。

[2] 根据主要设备放置的位置，空调系统可分为集中式，即整栋建筑作为一个整体进行集中调节，或分散式，即建筑根据调节特定区域不同进行分散调节。

[3] 集中式暖通空调系统可为一个或多个分区服务，其主要设备位于所服务分区外的合适位置，在建筑物内部、顶部或是邻近建筑物。

[4] 变风量系统包括一个中央空气处理机组，它向位于每个区域的变风量控制终端送风，以调节送风量。

[5] 建筑物需要同时进行分区供热和降温的，热泵将热量从一区域重新分配到另一区域，无须锅炉或冷却塔运行。

[6] 传热过程主要是通过居住者与辐射板之间的辐射方式，以及空气与辐射板之间的自然对流方式来实现的。

Extensive Reading

Air Conditioning Psychrometric

In air-conditioning system, the air must undergo one or several of the following processes. Psychrometrics, the science of moist air conditions, i.e. the characteristics of mixed air and water vapour, can be used to predict changes in the environment when the amount of heat and/or moisture in the air changes. Use of

Psychrometric Chart

psychrometric analysis is also important to determine the volume flow rates of air to be pushed into the ducting system and the sizing of the major system components.

Estimating Cooling & Heating Load

Load estimates are the summation of heat transfer elements into (gains) or out of (losses) the spaces of a building. Heat transfer elements are called load components, which can be assembled into one of three basic groups, external space loads, internal space loads and system loads. To properly understand the workings of the various external, internal and system load components, a series items need to be gathered from a set of plans, existing building surveys or occupant interviews, such as building square, year-round weather data (design conditions, heat transfer), construction materials [gather densities, external colour and U-factors or describe material type layer by layer (R-values)], ventilation needs (IAQ and exhaust makeup), etc. The total cooling load is then determined in kW or tons by the summation of all of the calculated heat gains. Along with psychrometrics, load estimating establishes the foundation upon which HVAC system design and operation occur.

Determine Design Supply Airflow Rate

HVAC engineers use psychrometrics to translate the knowledge of heating or cooling loads (which are in kW or tons) into volume flow rates (in m^3/s or CFM) for the air to be circulated into the duct system. The volume flow rate is used to determine the size of fans, grills, outlets, air-handling units, and packaged units. This in turn affects the physical size (foot print) of air handling units and package units and is the single most important factor in conceptualizing the space requirements for mechanical rooms and also the air-distribution ducts. The main function of the psychrometric analysis of an air-conditioning system is to determine the volume flow rates of air to be pushed into the ducting system and the sizing of the major system components.

Psychrometric Chart

The psychrometric chart provides a graphic relationship of the state or condition of the air at any particular time(Fig. 6-3). It displays the properties of air: dry bulb temperature (vertical lines), wet bulb temperature (lines sloping gently downward to the right), dew point temperature (horizontal lines), and relative humidity (the curves on the chart). Given any two of these properties, the other two can be determined using the chart. The chart's usefulness lies beyond the mere representation of these elementary properties—it also describes the air's moisture content (far right scale), energy content (outer diagonal scale on upper left), specific volume (lines sloping sharply downward to the right), and more.

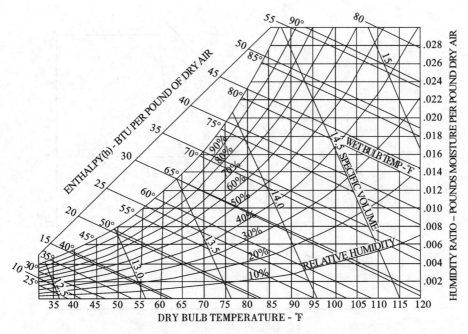

Figure 6-3 Psychrometric chart

Application of Psychrometric Chart: Cooling & Dehumidifying

When air is cooled below the dew point temperature, $T_{dewpoint}$, condensation occurs and moisture is removed from the air stream. The exiting air stream is at a lower temperature and humidity ratio than the incoming air stream. The cooling to condense water from the air is called latent cooling or dehumidification. Thus, this process includes both sensible and latent cooling. The total cooling is the sum of the latent and sensible cooling. The process is shown schematically below (Fig. 6-4).

On psychrometric chart, this process is represented as line sloping downward and to the left. This process is assumed to occur as simple cooling first and then condensation. While the moisture is condensing, the air is assumed to remain saturated. This process is used in air-conditioning systems operating in hot, humid climates. It is accomplished by using a cooling coil with a surface temperature below the dew point temperature of water vapor in air. Typical cooling coils in air conditioning systems operate at approximately 40-50° F, below the dew point temperature of typical indoor air conditions. The calculations are identical to those for heating and humidifying with the only difference being that the initial state (state 1) is the warmer, more humid state. As before, the total heat change (Q or q) in going from the initial to the final condition can be broken into a sensible and latent heat portion. This process is used in air-conditioning systems operating in hot, humid climates. Typical cooling and dehumidifying process shall include chilled

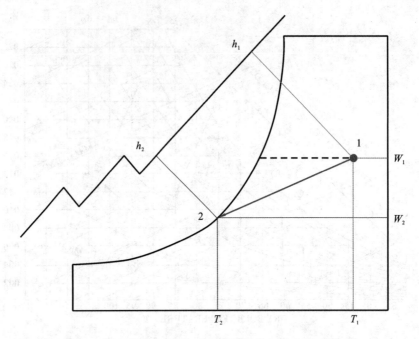

Figure 6-4 Psychrometric process: cooling & dehumidifying

water and refrigerant cooling coils which condition re-circulated room air or mixtures of re-circulated air and outdoor air which is introduced for ventilation. The cooling coil shall have a surface temperature below the dew point temperature of water vapor in air for effective condensation.

Example

Consider a hot humid day 90°F and 90% RH. We want to condition the air to 70°F at about 50% RH. We do this by chilling the air far enough to condense out enough moisture to dehumidify it; The goal is to have air with absolute humidity not exceeding 0.008 lbs of moisture per pound of air (50 to 55 grains per pound of dry air). Show the processes on the psychrometric chart(Fig. 6-5).

Solution

1. Plot 90°F and 90% RH on the chart.

2. Read the absolute humidity at this point to 0.029 lbs moisture per lb of dry air or (195 grains of moisture per pound of air).

3. Check the final condition of 0.008 lbs moisture per lb of dry air and run the horizontal line to saturation curve.

4. Read the temperature as 50°F.

5. Cool the air from 90°F at 90% RH to 50°F —Now there are 0.008 lbs moisture per lb of dry air and 100% RH. 0.021 lbs of moisture per lb of dry air (142 grains of moisture) have condensed out—The air is now dehumidified.

6. The air is dehumidified, but cold (50°F) and is at 100% RH; however it only has 53 grains of moisture.

7. Warm back up to 70°F (sensible heating), the RH raises to 50%.

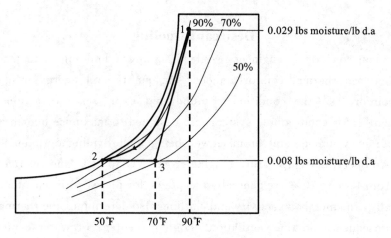

Figure 6-5 An example of psychrometric process

▶▶▶ Post-Reading Exercises

1. Choose the best answer for each of the following.

(1) Which are the influencing factors on the cooling/heating load? _____.

    A. Weather

    B. Construction material

    C. Occupants

(2) Which is the most important factor in conceptualizing the space requirements for mechanical rooms? _____.

    A. The size of fans, grills and outlets

    B. The volume flow rate

    C. The air distribution ducts

2. Translation.

(1) Psychrometrics, the science of moist air conditions, i.e. the characteristics of mixed air and water vapour, can be used to predict changes in the environment when the amount of heat and/or moisture in the air changes.

(2) Typical cooling coils in air conditioning systems operate at approximately 40-50°F, below the dew point temperature of typical indoor air conditions.

(3) The calculations are identical to those for heating and humidifying with the only difference being that the initial state (state 1) is the warmer, more humid state.

Desiccant Cooling

The main concern of engineers while designing a building is to maintain the balance between thermal comfort, indoor air quality and energy usage. CIBSE defines comfort as "the condition of mind that expresses satisfaction with the environment". Air-conditioning systems are designed to maximize human comfort in the interior environment and promise well-being by providing optimum indoor air quality. An operating temperature ranging from 18℃ to 26℃ and relative humidity ranging from 40% to 70% are generally acceptable for places with sedentary activity. Additionally, the metabolic activity and clothing also determines the thermal comfort and its consequence on the ventilation. High percentage of water content in the interior space can also give rise to various problems, they are:

1. Condensation on internal surfaces, which promotes mould growth and thus can be a source of several health problems.

2. Moisture can also lead to the corrosion of metal, decay of timber, and thus damage the internal structure.

Moisture content or latent heat of air can be controlled either by condensing the water vapour or by using suitable absorbents as used in desiccant cooling systems. While conventional VCSs simultaneously cool and dehumidify the air, a desiccant system only dehumidifies it. Moreover, a desiccant system can be used in combination with evaporative cooling system to maintain the temperature and moisture of incoming air. Earlier, desiccant systems were used for industrial and agricultural sector like textile mills, post-harvest crop storage units for humidity control and drying. However, energy crisis and necessity to develop more eco-friendly systems have led to the introduction of desiccant cooling systems as an effective method to control humidity.

Desiccant systems can use solid desiccants or liquid desiccants. Some commonly used solid desiccants include activated silica gel, titanium silicates, alumina, zeolite (natural and synthetic), molecular sieves, etc., whereas liquid desiccants comprise lithium chloride, lithium bromide, tri-ethylene glycol, calcium chloride and potassium formate. Apart from aforementioned desiccants, there are organic-based desiccants, polymeric desiccants, compound desiccants and composite desiccants. Desiccant systems include rotating desiccant wheel, solid packed tower, liquid spray tower, falling film and multiple vertical bed. Desiccant systems can be categorized

based on the type of desiccant used:
- Liquid desiccant systems,
- Solid desiccant systems,
- Advanced desiccants which include polymeric desiccant, composite desiccant, bio-desiccant.

Hybrid of LDS with Various Systems

Liquid Desiccant Combined with Evaporative Cooling System

The desiccant cooling system combined with an evaporative system used by Yin and his colleagues to perform an experimental study on the dehumidifier and regenerator. The system has three major components namely dehumidifier, regenerator and an evaporative cooler. Air after being dehumidified goes to the evaporative cooling unit, where the sensible heat is removed and introduced into the interior space while the diluted desiccant solution goes to the regenerator unit for regeneration. A storage tank is also provided, which can store the regenerated desiccant solution to be used when low-grade heat is not available. They used packed tower structure for both the dehumidifier and regenerator. Kim and his colleagues investigated an LDS integrated with an evaporative cooling assisted 100% outdoor air system. They investigated the energy saving potential of the system and compared it with a conventional variable air volume system. TRNSYS 16 was used for simulation and results concluded that the hybrid consumes 51% less cooling energy compared with VAV system.

Liquid Desiccant System Combined with VCS

Vapour compression systems are the most common systems for sensible cooling. They have been in operation for more than two decades now. Most of these systems are advanced and require less energy than their predecessors require. An LDS integrated with a VCS can be highly efficient in space cooling. Henning and his fellow researchers studied the potential of solar energy use in desiccant cooling systems and concluded that desiccant system combined with conventional system can save up to 50% of primary energy.

Khalil investigated the potential of one such system, which is called hybrid-desiccant-assisted air conditioner. The COP of the system tested was 3.8. Total cooling capacity of the system was 6.15 kW, using 2.6 kW VCS. He used lithium chloride as liquid desiccant. Fig. 6-6 shows the schematic diagram of hybrid system proposed by Khalil. Strong solution from the tank is pumped and sprayed uniformly over the evaporator surface area. Process air to be dehumidified is passed through the evaporator in the cross-flow direction. The evaporator and desiccant help in simultaneously cooling and dehumidifying the process air while the diluted desiccant solution is collected in the weak solution tank. After that, the diluted solution is pumped to absorb heat from the heat exchanger that uses the waste heat rejected from the condenser of the VCS to pre-heat the diluted solution. A heating coil in the regenerator tank provides the additional heat required by the solution. The proposed system can attain yearly energy savings of 53% compared with a VCS with a reheat mechanism.

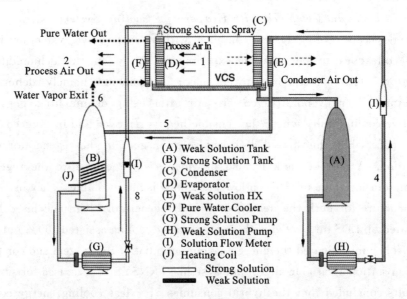

Figure 6-6　A schematic of hybrid desiccant-assisted air conditioner

A novel design of a hybrid system of VCS and LDS was proposed by She and his colleagues. In this system, liquid desiccant cooling system along with an indirect evaporative cooler was used to sub-cool the refrigerant of the liquid desiccant cooling cycle. Moreover, the desiccant solution was regenerated using the condensing heat of the VCS. Results obtained from the thermodynamic analysis showed that the proposed system attained higher coefficient of performance than conventional system as well as the reverse Carnot cycle under similar working conditions. About 18.6% and 16.3% higher COP were achieved using hot air and ambient air, respectively.

Liquid Desiccant-based Vapour Absorption System

Pang and his colleagues categorized absorption and adsorption system based on heat source and its application. The main function of both these systems is to refrigerate and dehumidify. Sarbu and his colleagues differentiated the two systems by their phenomenon. Absorption is a volumetric phenomenon, while adsorption is a surface phenomenon. A hybrid liquid desiccant-based air-conditioning system with a vapour absorption system was proposed by Ahmed and his colleagues. The hybrid is an open cycle system and used lithium bromide for both absorption and dehumidification. The simulation study showed that COP of the proposed hybrid system ranges from 0.96 to 1.25 and is 50% higher than the conventional vapour absorption system. In addition, they concluded that lower water temperature and moisture content would improve the performance further.

Post-Reading Exercises

1. Fill in the blanks.

(1) The definition of comfort is _____.

(2) The comfortable temperature range is _____ and relative humidity is _____.

(3) In liquid desiccant combined with evaporative cooling system, the function of evaporative cooling unit is to remove _____.

(4) In Khalil's study, the system's COP is _____. Strong solution is sprayed over _____.

(5) In hybrid system of VCS and LDS, liquid desiccant is used _____ _____, and regeneration of the desiccant solution is achieved by _____ _____.

2. Rephrase the following paragraph, without changing its meaning.

The desiccant cooling system has three major components namely dehumidifier, regenerator and an evaporative cooler. Air after being dehumidified goes to the evaporative cooling unit, where the sensible heat is removed and introduced into the interior space while the diluted desiccant solution goes to the regenerator unit for regeneration. A storage tank is also provided, which can store the regenerated desiccant solution to be used when low-grade heat is not available. They used packed tower structure for both the dehumidifier and regenerator. Kim and his colleagues investigated an LDS integrated with an evaporative cooling assisted 100% outdoor air system.

3. Choose the best answer for each of the following.

(1) The desiccant system can achieve _____.
  A. temperature control
  B. humidity control
  C. air quality control

(2) A desiccant system can be made of _____.
  A. silica gel

B. titanium silicates
C. zeolite
(3) The process air is _____ when passing through the evaporator.
A. heated
B. cooled
C. dehumidified

Part III
Building Water Supply and Drainage

Unit 7 Building Water Supply

Warm-up Activities

1. Learn the following words and find relevant information.

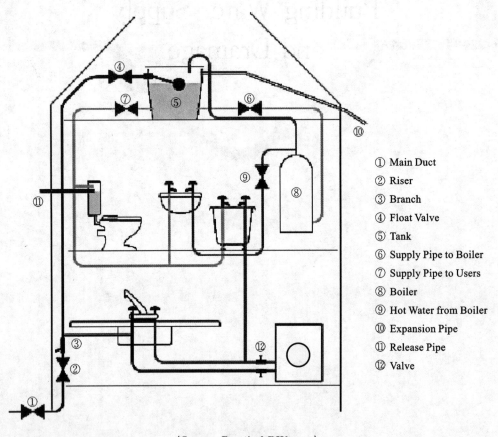

① Main Duct
② Riser
③ Branch
④ Float Valve
⑤ Tank
⑥ Supply Pipe to Boiler
⑦ Supply Pipe to Users
⑧ Boiler
⑨ Hot Water from Boiler
⑩ Expansion Pipe
⑪ Release Pipe
⑫ Valve

(Source: Practical DIY. com)

2. Please read the following paragraph and work through the relevant exercises.

Water Treatment

The type and extent of treatment required to obtain potable water depends on the quality of the source. The better the quality is, the less treatment is needed. Surface water usually needs more extensive treatment than groundwater does, because most streams, rivers, and lakes are polluted to some extent. Even in areas remote from human populations, surface water contains suspended silt, organic material, decaying vegetation, and microbes from animal wastes. Groundwater, on the other hand, is usually free of microbes and suspended solids because of natural filtration as the water moves through soil, though it often contains relatively highly concentrations of dissolved minerals from its direct contact with soil and rock.

Water is treated in a variety of physical and chemical methods. Treating of surface water begins with intake screens to protect fish and debris from entering the plant and damaging pumps and other components. Conventional treatment of water primarily involves clarification and disinfection. Clarification removes the most of the turbidity, making the water crystal clear. Disinfection, usually the final step in the treatment of drinking water, destroys pathogenic microbes. Groundwater does not often need clarification, but it should be disinfected as a precaution to protect public health. In addition to clarification and disinfection, the processes of softening, aeration, carbon adsorption, and fluoridation may be used to certain public water sources. Desalination processes are used in areas where freshwater supplies are not readily available.

(1) Correct the grammar errors in the following sentences.

ⅰ. Even in areas remote from human populations, surface water contain suspended silt, organic material, decaying vegetation, and microbes from animal wastes.

ⅱ. Groundwater, on other hand, is usually free of microbes and suspended solids because of natural filtration as the water moves through soil, though it often contains relatively highly concentrations of dissolved minerals from its direct contact with soil and rock.

ⅲ. Treating of surface water begins with intake screens to protecting fish and debris from entering the plant and damaging pumps and other components.

ⅳ. Clarification removes the most of the turbidity, to make the water crystal clear.

ⅴ. In addition clarification and disinfection, the processes of softening, aeration, carbon adsorption, and fluoridation may be used to certain public water sources.

(2) Oral exercise: Brief your classmate the steps shown in the diagram below.

(Source: Denver Water)

Introduction to Water Supply

cistern
蓄水池

draw-off point
引出点

Intensive Reading

General Water Supply

Water supply is the provision of water by public utilities, commercial organizations, community endeavours or by individuals, usually via a system of pumps and pipes[1]. It usually includes hot and cold-water distribution system that are supplied throughout the entire building, the means of pumping or applying the water main to maintain the water pressure in order to assure that the highest outlet has an adequate water flow during crucial time.

Cold Water Services

Water may be supplied to cold taps either directly from the mains via the supply pipe or indirectly from a protected cold-water storage cistern. In some cases, a combination of both methods of supply may be the best arrangement.

A supply "direct" from the mains is preferred because water quality from storage cannot be guaranteed. However, pressure reliability of the mains supply should be considered especially where connections are made near to the ends of distributing mains [2]. Where constant supply pressure may be a problem, storage should be considered.

Factors to consider when designing a cold-water system should be taken into account, such as the available pressure and reliability of supply, particularly where any draw-off point is at the extreme end of a supply pipe or situated near the limit of main pressure.

For large buildings, such as office blocks, hostels and factories, it will usually be preferable for all water, except drinking water, to be supplied indirectly from cold water storage cisterns. Drinking water should be taken directly from the water supplier's main wherever practicable.

Service pipe diameter should be determined by hydraulic calculation after consultation with the consumer and the water supplier. This is particularly important in the case of nondomestic users where design requirements may create greater demands on the available supply [3]. Calculations should also take into account any future supply needs that might be expected. In premises where water for firefighting purposes is required, the local fire service should be consulted and relevant regulations complied with.

Pipelines should be designed to accommodate thermal movement. Particular attention should be given to the difference in temperature during installation compared to the expected temperature when they are filled and in use. Considerable stress can occur in pipelines where insufficient expansion joints are inserted. This can be a particular problem when plastics pipes are pulled through the ground by mole plough when stresses caused by "stretching" are combined with those of contraction through cooling.

Where the height of building lies above statutory level or when the available pressure is insufficient to supply the whole of a building and the water supplier is unable to increase the supply pressure in the supplier's mains, consideration should be given to the provision of a pump [4]. Where the pump delivers 0.12 L/s the water owner must be notified and written consent obtained before then pump is fitted. It is important that the water supplier is consulted and written consent received before fitting any pump. The water supplier will wish to ensure that water regulations are complied with respect of backflow risk, and that any pump will not have adverse effects on the mains and other users.

Materials for Water Services

Hot-and cold-water systems are normally semi-hard copper tube with compression, soldered or push-fit polybutylene connections. Galvanized and stainless-steel pipes and fittings are used in large installations such as industrial, hospital and campus sites. Lead pipes may still be found in old buildings or as underground water mains and these should be replaced. Galvanized steel pipes corrode and eventually block with rust deposits, discolouring and stopping water flow. Corrosion protection is provided by ensuring that incompatible materials such as copper and zinc galvanizing are not mixed in the same pipe system. Hot-and cold-water service systems are continually flushed with fresh water, making it necessary to use galvanized metal, copper or stainless steel as the water cannot be treated with chemicals. Copper and galvanized steel should not be used in the same system because electrolytic action will remove the internal zinc coating and cause pipe failure. A galvanized metal cold-water storage tank can be successfully used with

---

hydraulic calculation
水力计算

premise
前提

accommodate
适应

thermal movement
因热产生的移动

mole plough
鼹鼠犁

contraction
收缩

statutory level
法定水平

galvanized
镀锌

copper pipes as the low temperature in this region limits electrolytic action. Heat accelerates all corrosion activity.

Black mild steel is used in recirculatory heating systems. An initial layer of mill scale, which is metal oxide scale formed during the high-temperature working of the steel during its manufacture, helps to slow further corrosion. Discoloration of the central heating water from rust to black during use shows steady corrosion. Black metallic sludge forms at low points after some years. Large hot-water and steam systems have the mill scale chemically removed during commissioning and corrosion-inhibiting chemicals are mixed with the water to maintain cleanliness and avoid further deterioration. Methane gas forms in closed heating systems during the first year of use, due to early rapid corrosion and radiators need frequent venting to maintain water levels. Proprietary inhibitors should be added to all central heating systems. These control methods of corrosion are anti-bacterial. Without them, steel boilers and radiators can rust through in 10 years.

Hose Reel

Hose reels are firefighting equipment for use as a first-aid measure by building occupants. They should be located where users are least likely to be endangered by the fire, i.e. the staircase landing [5]. The hose most distant from the source of water should be capable of discharging 0.4 L/s at a 6 m distance from the nozzle, when the two most remote hose reels are operating simultaneously. A pressure of 200 kPa is required at the highest reel. If the water main cannot provide this, a break/suction tank and booster pumps should be installed. The tank must have a minimum volume of water of 1.6 m^3. A DN50 supply pipe is adequate for buildings up to 15 m height and a DN65 pipe will be sufficient for buildings greater than this. Fixed or swinging hose reels are located in wall recesses at a height of about 1 m above floor level. They are supplied by a DN25 pipe to DN20 reinforced non-kink rubber hose in lengths up to 45 m to cover 800 m^2 of floor area per installation(Fig. 7-1).

The specification of a sprinkler system will depend on the purpose intended for a building, its content, function, occupancy, size and disposition of rooms. Installations to commercial and industrial premises may be of the following types.

• Dry riser.

A dry riser is in effect an empty vertical pipe which becomes a fire fighter's hose extension to supply hydrants at each floor level. A dry riser is installed either in unheated buildings or where the water main will not provide sufficient pressure at the highest landing valve.

• Wet riser.

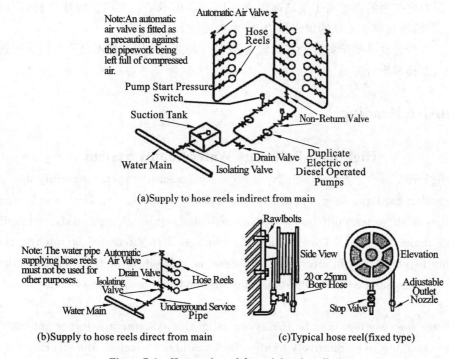

Figure 7-1　Hose reels and foam inlets installations

A wet riser is suitable in any building where hydrant installations are specified. It is essential in buildings where floor levels are higher than that served by a dry riser, i.e. greater than 60 m above fire service vehicle access level. A wet riser is constantly charged with water at a minimum running pressure of 400 kPa with up to three most remote landing valves operating simultaneously. A flow rate of 25 L/s is also required. The maximum pressure with one outlet open is 500 kPa to protect firefighting hoses from rupturing.

Pumping

To maintain water at the required pressure and delivery rate, it is usually necessary to install pumping equipment. Direct pumping from the main is unacceptable. A suction or break tank with a minimum water volume of 45 m^3 is used with duplicate power source service pumps. One 65 mm landing valve should be provided for every 900 m^2 floor area.

suction
抽吸
break tank
破碎罐

Notes

［1］供水系统是通过公共设施、商业机构、社区的努力或通过个人提供水的系统。水的输送通常采用水泵与管道系统。

［2］但是，在靠近主管的连接处应考虑水源的压力可靠性。

［3］对于非民用用水单位而言，当设计需求对水的供应有更高的要求时，这一点尤为重要。

[4]当建筑物的高度高于规定高度或可用的水压不足以供应整幢建筑物，而用水单位又不能增加供水总管的供水压力时，须考虑设置水泵。

[5]软管卷盘是用于急救的消防设备。它们应置于使用者最不可能受到火灾危害的地方，即楼梯平台。

Extensive Reading

High-Rise Buildings Water Supply System

High-rise buildings, especially multi-story structural systems mainly apply to office buildings, large shopping centres and exhibition halls, etc. High-rise buildings are often with water uplifting system, particularly through equipped intermediate storage tanks or basins. The mentioned tanks are distributed throughout the entire building height at certain intervals. The water is uplifted to the tank, located above, and later it is pumped from the tank and supplied to another tank, located above, and etc.

Since the very high-rise buildings are using this system, it is important that the system structure to be relatively light-weight would not overload the structure of the building with useless weight. The system would also be capable of treating water from the sea or other surface water pools and turn it into potable water or water suitable for daily activities. With the reference to these criteria, a water supply system has been developed for high-rise buildings, whose operation principle lies in water evaporation, which is performed at the lower part of the building, while water steam is uplifted in shafts being affected by natural traction. Upon reaching the top of the shaft water steam gets cool up to the temperature of dew point and condenses. Condensed water is collected and supplied to pipelines. There is also an option to direct surplus water downwards by pipelines equipped with turbines for power production.

Copper is the most common material for supply pipe; it resists corrosion, so water runs freely and pipes don't leak for many decades. It is often equipped with galvanized steel pipes in an older home, which tends to clog with minerals and rust over time and can develop leaks. Nowadays, many homes have plastic supply pipes, most commonly cross-linked polyethylene (PEX), which are installed quickly and are expected to last virtually forever.

Water arrives via a main supply pipe, which is typically 1 inch in diameter or larger. (A pipe is measured by its inside dimension, so 1-inch copper pipe is about 1.125 inches in outside diameter.) In most cases, the pipe runs through at least one main shutoff valve, located outside the house in a "Buffalo box" buried in the yard near the house or just inside the basement or crawl space. It then usually passes through a water meter, and there is likely another main shutoff after the water meter.

The main supply line usually runs to the water heater where it is divided into cold and hot water pipes. From there, supply pipes almost always travel in pairs, hot and cold. Pipes from the water heater are typically 3/4 inch but may be 1/2 inch. Horizontal pairs run to below walls and then vertical pairs, called risers, run up to various rooms.

In newer homes, there are separate lines running from the water heater to each room, so water use in one area does not affect use in another area. In an older home, single line may loop throughout the house, meaning, for instance, that if someone flushes a toilet downstairs, the cold water supplying a shower upstairs will have lessened pressure, causing the shower water to suddenly become hot.

This water supply system for high-rise and particularly high-rise buildings are used to the ones that are located in the regions of hot climate, where solar energy is sufficient for the evapouration of the required amount of water.

Basically, the high-rise building water supply system comprises of a water basin with casing, made of transparent material, the shaft for water steam flow and upper condensation facility, from which the water may flow into the upper tank. The mentioned water basin may be a land-based water tank, which is continually supplied with ground or surface (river, sea, etc.) water. The basin may also be equipped in the shore of a large water body (sea, ocean, river, lake, etc.), if the building is located close to such water body. In both cases, a transparent casing is arranged above the water surface, leaving an air gap. This air gap is used to transfer water steam affected by natural traction, which appears in the shaft due to temperature difference. In order to establish conditions under which a constant one direction steam flow would appear in the mentioned air gap, one side of the casing shall open for environmental air to pass through it. Thus, the casing at one side has an opening, at the other-contraction for water steam flow to pass to the lifting shaft. Casing flanks are closed and directly attached to the walls of the basin. In case the basin is arranged in the shore of a large water pool, the bottom of the basin is made inclined, a water outlet is left at the bottom of the inclination in basin's wall for water outflow, and therefore, the remaining water with the major salt and sewage content settles to the bottom and withdraws via the stated outlet. Water change of tide takes place, which ensures prevention of salt and other sewage sediments accumulation in the basin. The walls of the shaft are properly insulated from the outside in order that the steam temperature throughout the entire height of the shaft would drop as less as possible. By sustaining approximately constant temperature in one shaft, steam does not condense on the half-way and useless condensate does not accumulate.

Post-Reading Exercises

1. Translation.

(1) The mentioned tanks are distributed throughout the entire building height at certain intervals.

(2) The water is uplifted to the tank, located above, and later it is pumped from the tank and supplied to another tank, located above, and etc.

(3) Basically, the high-rise building water supply system comprises of a water basin with casing, made of transparent material, the shaft for water steam flow and upper condensation facility, from which the water may flow into the upper tank.

2. Fill in the blanks with the words or expressions given below, and change the form where necessary.

Since water does not flow _____, high rise buildings depend on pumps to "lift" the water to the upper floors. A typical municipal water system does not have enough _____ to do the job by itself. High rise buildings _____ chilled water and hot water air conditioning systems plus fire sprinklers have three systems. This means there are separate pumps for each system, usually located in mechanical rooms or sprinkler rooms. These pumps may be operated in circuits _____ additional pumps are required at higher floors, to maintain pressure on each system without _____ the primary pumps.

_____ on the design of the building, there may be mechanical or plumbing rooms _____ different levels with pressure tanks in _____ to regulate the water supply pressure to each system, decreasing _____ when pumps are running, and to maintain pressure when the pumps cycle _____.

| where | on | with | each | overload |
| off | head | uphill | overpressure | depend |

Pumps: the Tree Main Types

There are very many types of pumps. Broadly they fall into three main categories:
- Displacement pumps,
- Rotary pumps,
- Centrifugal pumps.

Pump

All these types are obtainable in sizes to meet almost every requirement, but each has its advantages and its limitations, and selection must take into consideration all aspects of the application —the height the water to be elevated, the cleanliness of the water, the form of power available to work the pump, the amount of noise that can be tolerated, and the length of life required of the mechanism.

Most hand-operated pumps are of the displacement type, having a plunger which is lifted and lowered alternately, or pushed backwards and forwards in a cylinder. Another variety, called the *semi-rotary pump*, has a metal plate which fits diametrically across a round chamber and is oscillated with a backward and forward semi-rotary movement; the plate is fitted with valves, so is the chamber, which is divided so that the pump has the equivalent of two cylinders. This is a popular type for raising water from a well to an overhead tank. The plungers in displacement pumps are fitted with cups made of leather or fabric, or they operate through packed glands. Usually these pumps give many years of reliable service; the cups or packing are renewed at intervals and the material costs are low.

Rotary pumps of the gear type are convenient and inexpensive, but rely on close metal-to-metal clearances, and if the water they handle is gritty they wear quickly and become noisy and inefficient. They are particularly useful for pumping oil and chemicals, where cleanliness is assured. They usually rotate at speeds of 1000 rpm or more.

Some metal-and-rubber rotaries are useful and reliable for small water-supply duties and have good self-priming characteristics. These pumps usually run at 1450 rpm —a useful electric motor speed—and are quiet in operation.

As both displacement and rotary pumps are positive in action, they give approximately constant volume at any fixed speed, irrespective of the pressure.

Centrifugal pumps usually run at 1450 or 2900 rpm and are often direct-coupled to electric motors. Ordinary centrifugal pumps have no ability to self-prime; the pipe system must be completely filled with water and all air must be released before the pump is started. But there are several devices which can be incorporated in the design to provide self-priming; most include circulation of part of the water from the delivery side, back to the suction side through a Venturi tube. This draws water up the suction pipe and dispels the air in the system.

Centrifugal pumps are not positive in action and the volume delivered varies widely at different pressures, even at constant speed. Most of these centrifugal pumps can draw up water 4.5-6.0 m vertically. The total head these pumps can produce depends on the peripheral speed of the impeller. Most small single-impeller pumps give total heads of 10-20 m. Larger single-stage pumps give up to 60 m. To attain greater heads several pump units are assembled in series on one spindle and are driven by the same motor.

Reciprocating displacement pump has its cylinder located at or below water level, suspended from ground level. Its operating rod is extended above ground and worked by crank mechanism driven by an engine, electric motor or windmill.

》》》 Post-Reading Exercises

1. Choose the best answer for each of the following.

(1) Most hand-operated pumps are _____.
   A. displacement pumps
   B. rotary pumps
   C. centrifugal pumps

(2) Which of following statements is true? _____.
   A. Most of these centrifugal pumps can draw up water 4.5-6.0 m vertically
   B. Reciprocating displacement pump has its cylinder located above water level
   C. Rotary pumps rotate at 2900 rpm

(3) Which of following pumps is useful for pumping oil and chemicals? _____.
   A. Displacement pumps
   B. Rotary pumps
   C. Centrifugal pumps

(4) Which of following pumps is a popular type for raising water from a well to an overhead tank? _____.
   A. Displacement pumps
   B. Semi-rotary pumps
   C. Centrifugal pumps

(5) Most centrifugal pumps can draw up water _____ vertically.
   A. 60 m
   B. 4.5-6 m
   C. 10-20 m

2. Short answer.

(1) What is self-priming?

(2) What is the characteristics of centrifugal pumps?

(3) When selecting a pump, what are the parameters we must take into consideration?

Unit 8　Building Drainage System

Warm-up Activities

1. Learn the following words and find relevant information.

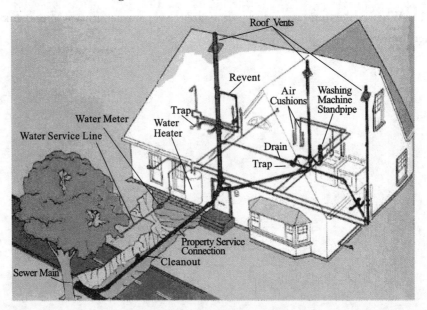

(Source: MSD Water Services)

2. The following are the steps in the design process for a drainage system on a standard building. Please place them in right order.

ANS: The right order is _____.

a. Determine and check the catchment area per downpipe. The area of round pipes should equal to the area of the gutter. However, square or rectangular pipes may be up to 20% smaller.

b. Determine gutter outlets, such as sumps, rain heads, and nozzles.

c. Determine the area for your proposed eaves gutter.

d. Determine the average recurrence interval (ARI) of rainfall in the area. This is a year, generally between 20 and 100.

e. Determine the catchment area with the slope.

f. Determine the downpipe size. The area of round pipes should equal the area of the gutter. However, square or rectangular pipes may be up to 20% smaller.

 g. Determine the gutter size, capacity and fall. To prolong the life of the gutter, you should install them with a steep fall, which allows water to easily flow away and properly drain.

 h. Determine the number of downpipes required, where they should be located, and what the high points are.

 i. Establish overflow measures and considerations.

 j. Measure the roof dimensions.

 k. Obtain the rainfall intensity of the site.

 3. Translation.

（1）A building drainage system used for apartment houses, detached houses, etc. generally includes drainage stack passing through stories.

（2）A discharge path is formed from the reservoir to the surrounding soil or through a pipe to a remote located pop-up discharge valve.

Intensive Reading

Building Drainage System

 Drainage is the method of removing surface or sub-surface water from a given area. Drainage systems include all of the piping within a private or public property that conveys sewage, rainwater, and other liquid wastes to a point of disposal [1]. The main objective of a drainage system is to collect and remove waste matter systematically to maintain healthy conditions in a building. Drainage systems are designed to dispose of wastewater as quickly as possible and should prevent gases from sewers and septic tanks from entering residential areas.

 Drainage systems can be classified into the sanitary sewer system and storm sewer system.

 The sanitary sewer is a system of underground pipes that carries sewage from bathrooms, sinks, kitchens, and other plumbing components to a wastewater treatment plant where it is filtered, treated and discharged [2]. The storm sewer is a system designed to carry rainfall runoff and other drainage [3]. It is not designed to carry sewage or accept hazardous wastes. The runoff is carried in underground

sanitary sewer
下水道
storm sewer
雨水管
rainfall runoff
降雨径流
hazardous waste
危险废物

pipes or open ditches and discharges untreated into local streams, rivers and other surface water bodies. Storm drain inlets are typically found in curbs and low-lying outdoor areas. Some older buildings have basement floor drains that connect to the storm sewer system.

Sanitary Sewer System

Sanitary pipework is aboveground pipework that is used to carry water from toilets, sinks, basins, baths, showers, bidets, dishwashers, washing machines, and so on, out of a building to the sewage system. The equivalent underground pipework is referred to as foul drainage and sewers [4]. According to the Building Regulations in UK, every modern home must have a WC fitted that is connected direct to the drainage system, with a basin fitted next to the WC with a supply of hot and cold water. In addition, every home must have a fixed bath or shower with a hot and cold water supply, and all appliance connections that are connected to the drainage system must have a trap to prevent odours and dangerous gases from building up inside.

Terminology of drain systems includes:

• Combined system in which foul and surface-water are conveyed in the same pipe.

• Discharge stack is vertical pipe conveying foul fluid/solid.

• Foul drain conveying black waste water material.

• Sewer pipe system provided by the local drainage authority.

• Invert is the lowest point on the internal surface of a pipe.

• Separate system in which foul and surface-water are discharged into separate sewers or places of disposal.

• Waste pipe from a sanitary appliance to a stack.

Traditionally, sanitary pipework was manufactured using metals such as cast iron, copper or lead; however, modern designs predominantly use plastics such as uPVC, high-density polyethylene (HDPE), polypropylene and so on. They are generally connected by some form of welding (such as solvent welding) or using push-fit fittings. The pipes are designed to different diameters depending on the appliance which they are connected. For example, a WC typically uses a 110 mm diameter pipe; baths, showers, sinks, washing machines use 40 mm diameter pipes; a bidet uses a 32 mm diameter pipe, and so on. The pipes are laid out to a slope or "fall" which allows wastewater to drain away without leaving debris behind and avoiding blockages. Discharge from a sanitary appliance into drain pipes is a random occurrence of surges of fluid [5]. The pipe flows full at some time and a partially evacuated space appears between the remaining trap seal and the surging fluid.

Separation between the water attempting to remain in the P-trap and the plug falling into the soil stack causes an air pocket to form. The static pressure of this air will be sub-atmospheric. Air from the room and the ventilated soil stack bubbles through the water to equalize the pressures and a noisy appliance operation result. The inertia of the discharge may be sufficient to syphon most of the water away from the trap, leaving an inadequate or non-existent seal. The problem is avoided by using 32 mm basin waste pipes when the length is restricted to 1.7 m at a slope of 20 mm/m run.

sub-atmospheric
亚大气

Sanitary pipework is typically connected to a soil vent pipe(SVP) which is a vertical pipe often attached to the exterior of a building that connects the drainage system to a point just above the level of the roof gutter allowing odours to be released. Venting the stack prevents water seals in traps from being broken by pressure buildups in the system. The sloping waste pipe can be up to 3 m long if its diameter is raised to 40 mm after the first 50 mm of run. This allows aeration from the stack along the top of the sloping section. Longer waste pipes with bends and steeper or even vertical parts have a 25 mm open vent pipe. Vertical soil and vent stacks are open to the atmosphere 900 mm above the top of any window or roof-light within 3 m. Underground foul sewers are atmospherically ventilated. Water discharged into the stack from an appliance entrains air downwards and establishes air flow rates up to a hundred times of the water volume flow rate. Air flow rates of 10 to 150 L/s have been measured. The action of water sucking air into the pipe lowers the air static pressure which is further reduced by friction losses. Water enters the stack as a full-bore jet, shooting across to the opposite wall, falling and establishing a downward helical layer attached to the pipe surface. Restricted air passageways at such junctions further lower the air static pressure.

soil vent pipe
排土管

Storm Sewer System

It is a requirement of Building Regulation's Approved Document H that adequate provision is made for rainwater to be carried from the roof of buildings. To achieve this, roofs must be designed with a suitable fall towards either a surface water collection channel or gutter that conveys surface water to vertical rainwater pipes, which in turn connect the discharge to the drainage system. The type of roof covering used determines the required fall of the roof. Minimum recommended falls are typically:

- Aluminium — 1∶60.
- Lead — 1∶120.
- Copper — 1∶60.
- Roofing felts — 1∶60.
- Mastic asphalt — 1∶80.

Roof Drainage System

Standard "flat" roofs should have a designed minimum fall of 1∶40, so that an actual finished fall of 1∶80 is achieved, allowing some room for error in the construction [6]. Drainage from roofs is generally provided by internal rainwater outlets and downpipes, or by external guttering systems or hoppers. It is recommended that there are at least two drainage points, even if the roof is small, to mitigate against one of them becoming blocked.

Interior Drains

Interior drains work just like the drain in your shower or sink. These drains are placed in areas of the roof that collect the most water, and they lead the water into a system of pipes that is installed below the roof. The water travels through these pipes until it is dispensed into a gutter or downspout at the side of the building.

Gutters

Gutters are the most commonly used, and most cost-effective drainage solution for flat roofs. They catch rainwater as it rolls off the edge of the roof and divert the water into a downspout that dispenses it a safe distance from the foundation of the building. This prevents the water from rolling off the roof uncontrolled and running down the side of the building which could damage the siding, windows, and foundation. There are a couple of disadvantages to using gutters on flat roofs. Gutters need consistent cleaning throughout the year because they gather debris that can block the flow of the water. If this debris is not cleared out, the water will overflow and run down the side of the building. Gutters are also susceptible to damage from severe weather, ice, and heavy debris. It is important to weigh these disadvantages against the low cost of gutters before making a final decision.

Scuppers

Scuppers are the most effective drainage solution for flat roofs. With this system, large square openings are made along the edge of the roof that shoot the water away from the side of the building. Sometimes downspouts are installed directly below these openings to catch the water and drain it away from the building and foundation in a controlled manner.

Flat roof scuppers have the following benefits:

- Cost effective.
- Easy to maintain.
- Large, wide scuppers rarely if ever get clogged by debris.

Notes

[1]排水系统包括所有私人或公共建筑中的管道。这部分管道将污水、雨水和其他液体废物输送至处理点。

[2]污水管系统利用一个地下管道将浴室、水槽、厨房和其他管道部件的污水输送

到污水处理厂进行过滤、处理和排放。

［3］雨水管系统是用来输送降雨径流和其他排水的系统。

［4］与之相对应的地下管道工程称为污水排放下水道。

［5］从卫生器具排放到排水管中的液体是冲击流,而且发生时间不定。

［6］标准"平"屋顶应按照最小设计落差1∶40进行设计,鉴于施工中存在一定的误差,实际完工后落差可达到1∶80。

Extensive Reading

Accessories on Water Supply and Drainage Systems

Joints

Rigid joints of clay drain pipes are now rarely specified as flexible joints have significant advantages:

- They are quicker and simpler to make.
- The pipeline can be tested immediately.
- There is no delay in joint setting due to the weather.
- They absorb ground movement and vibration without fracturing the pipe.

Modern pipe manufacturers have produced their own variations on flexible joints, most using plain ended pipes with a polypropylene sleeve coupling containing a sealing ring, as shown in Fig. 8-1. Cast iron pipes can have spigots and sockets caulked with lead wool. Alternatively, the pipe can be produced with plain ends and jointed by rubber sleeve and two bolted couplings. Spigot and socket uPVC pipes may be jointed by solvent cement or with a push-fit rubber "O" ring seal. They may also have plain ends jointed with a uPVC sleeve coupling containing a sealing ring.

Fittings: Traps

Foul air from the drain and sewer is prevented from penetrating buildings by applying a water trap to all sanitary appliances. A water seal trap is an integral part of gullies and WCs, being mounded in during manufacture. Smaller fittings, i.e. sinks, basins, etc., must be fitted with a trap. The format of a traditional tubular trap follows the outline of the letter "P" or "S"(Fig. 8-2).

Apparatus: Automatic Flushing Cisterns

Flushing cistern is used for automatically flushing WCs. It has application to children's lavatories and other situations where the users are unable to operate a manual flush device. As the cistern fills, air in the stand pipe is gradually compressed. When the head of water "H" is slightly above the head of water "h", water in the trap is forced out. Siphonic action is established and the cistern flushes the WC until air enters under the dome to break the siphon.

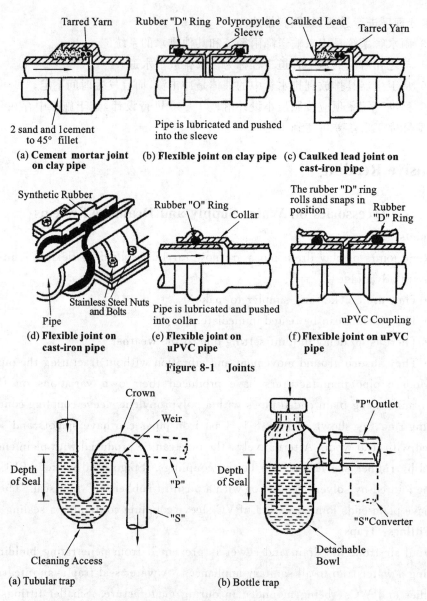

Figure 8-1　Joints

Figure 8-2　Traps

With the smaller urinal flush cistern, water rises inside the cistern until it reaches an air hole. Air inside the dome is trapped and compressed as the water rises. When water rises above the dome, compressed air forces water out of the U tube. This lowers the air pressure in the stand pipe creating a siphon to empty the cistern. Water in the reserve chamber is siphoned through the siphon tube to the lower well.

Apparatus: Flushing Valves

Flushing valves are a more compact alternative to flushing cisterns, often used in marine applications, but may only be used in buildings with approval of the local water authority. The device is a large equilibrium valve that can be flushed at any time without delay, provided there is a constant source of water from a storage cistern. The minimum and maximum head of water above valves is 2.2 m and 36 m respectively. When the flushing handle is operated, the release valve is tilted and water displaced from the upper chamber. The greater force of water under piston "A" lifts valve "B" from its seating and water flows through the outlet. Water flows through the by-pass and refills the upper chamber to cancel out the upward force acting under piston "A". Valve "B" closes under its own weight(Fig. 8-3).

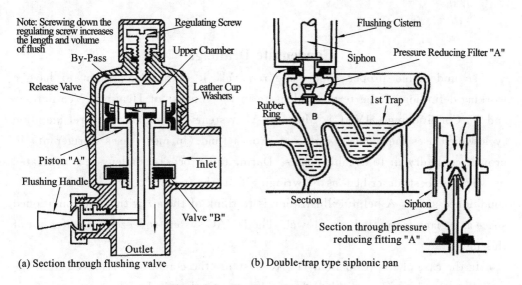

Figure 8-3 Flushing valve & cisterns

>>> Post-Reading Exercises

1. True or False.

(1) Flushing valves are a more compact alternative to flushing cisterns. (   )

(2) Rigid joints of clay drain pipes are now specified as flexible joints. (   )

(3) When the head of water "H" is slightly above the head of water "h", water in the trap is forced out.   (   )

(4) Foul air from the drain and sewer is prevented from penetrating buildings by applying a water trap to all sanitary appliances.   (   )

(5) Cast iron pipes can have spigots and sockets not caulked with lead wool.

(   )

2. Translation.

(1) With the smaller urinal flush cistern, water rises inside the cistern until it

reaches an air hole. Air inside the dome is trapped and compressed as the water rises.

(2) The device is a large equilibrium valve that can be flushed at any time without delay, provided there is a constant source of water from a storage cistern.

(3) They absorb ground movement and vibration without fracturing the pipe.

Condensate Drainage

To understand proper condensate removal, it is first important to have a working definition of the term condensate in the context of Heating, Ventilation, and Air Conditioning (HVAC). Most HVAC systems contain a form of refrigeration cycle which uses mechanical work acting on a fluid (in most cases refrigerant) to provide cool, dry air to liveable spaces. During this process, condensate is generated by what is best described as, "wringing out the water", from the occupied/conditioned space. A helpful illustration is to think of the air within the conditioned space as a giant, saturated, wet towel. The HVAC system squeezes the water out of this towel.

In the case of the residential HVAC system, the condensate removed from the air falls to a drain pan within what is known as the air handling unit. The air handling unit (sometimes furnace) is the indoor portion of a split system and is often located in an attic space or within a basement. Contained within the air handling unit is the coil which accumulates the condensate and it is then disposed of as water in liquid form. The drawing detail below identifies the main components of the air handling unit(Fig. 8-4):

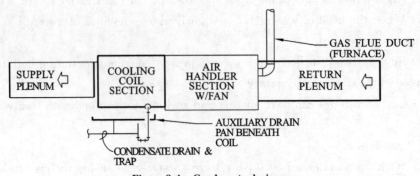

Figure 8-4 Condensate drainage

The conveyance and/or disposal of the water from the air handling unit is often a point of failure due to improper installation, lack of maintenance, or exposure to freezing temperatures. Some key points are identified below from the International Residential Code to address installation defects:
- Condensate from all cooling coils or evaporators shall be conveyed from the drain pan outlet to an approved place of disposal.
  ◦ It is important to consult local authority regarding approved disposal locations. Different municipalities may require that condensate be disposed of to the sanitary sewer, while others may require disposal to building exterior or storm drainage piping. Some AHJ may even require the use of additional equipment to neutralize certain elements/compounds prior to disposing of the condensate.
- Horizontal slope must be no less than 1/8 unit vertical in 12 units horizontal.
  ◦ Many homeowners experience an unintended water discharge from an air handling unit located in an attic space because the installing contractor does not provide adequate "fall" to the condensate drain piping to permit gravity drainage. This is considered a defect in installation.
- Condensate shall not discharge to a street, alley, or other areas where it would cause a nuisance.
  ◦ Some installations discharge condensate to areas where there may be pedestrian foot traffic. If condensate is discharged to a walkway, it may create a slipping hazard.

P-trap installation can be a source of improper installation. P-traps for use in HVAC applications vary in design, however, the correct trap depends on both the air handling unit's components as well as the air distribution system (ductwork). The P-trap must always contain the required amount of water to prevent contaminants from entering the HVAC system. In many residential applications, the duct system may have many bends, transitions, or flex-connections where the supply fan of the air handling unit must overcome the added pressure. If the static pressure of the duct system is high, the water-seal of the P-trap may be "pushed" out, allowing contaminants to enter the duct system. Consult with the equipment manufacturer's data to ensure the proper drain piping and P-trap is used for the system.

A third and common source of improper installation can be found in the secondary (emergency) drainage system. A secondary drainage system is required where the threat of an overflow may damage building components and is often accomplished by a "secondary drain pan" installed under the air handling unit. The secondary drainage system must provide a method, typically a moisture-sensing switch, to shut down power to the air handling unit, in the event moisture is detected. Many secondary drainage defects include moisture sensing switch faults, improper secondary drain pan piping, or in some instances, no secondary drain pan may be installed at all.

With the increased popularity of high-efficiency equipment, it is important to be aware that these systems can produce condensate year-round, including during the winter months. Installation contractors may plumb the condensate drain to discharge to the outside, as is their usual practice. However, in the case of a high-efficiency furnace, condensate can form in the exhaust gases when the unit is in heating mode. The condensate will then drain to the outside where it is exposed to freezing temperatures, resulting in a backup. Safety switches within the air handling unit are intended to shut the unit off when the condensate backs up. This scenario also applies to the high-efficiency water heaters and the increased popularity of instantaneous water heaters.

Proper maintenance will aid in preventing drainage system failures. Typical maintenance for a condensate drainage system consists of a yearly inspection and in some cases, detergent cleaning of the system. Cleaning is required due to the occasional build-up of debris and material which can accumulate within the drains.

In closing, the conveyance and/or disposal of the water from the air handling unit is often a point of failure related to improper installation, lack of maintenance, or exposure to freezing temperatures. Correct installation, as well as proper maintenance, will address many potential issues.

Post-Reading Exercises

1. Translation.

(1) Condensate from all cooling coils or evaporators shall be conveyed from the drain pan outlet to an approved place of disposal.

(2) Many homeowners experience an unintended water discharge from an air handling unit located in an attic space because the installing contractor did not provide adequate "fall" to the condensate drain piping to permit gravity drainage. This is considered a defect in installation.

(3) Condensate shall not discharge to a street, alley, or other areas where it

would cause a nuisance.

2. Fill in the blanks with the words or expressions given below, and change the form where necessary.

The water might be condensate coming from your air conditioning system _____ is probably what you shall worry about when you first noticed the moisture.

Condensate is _____ water vapor, which is a _____ of your cooling system; it is the moisture that AC system _____ from the air and "condenses" into liquid form in the process of cooling the air in your space. Normally, condensate _____ on the evaporator coil and _____ away through a tube called a drain line, sometimes _____ into a drain pan, and sometimes pumped away by a condensate pump, _____ on the design of your system.

So there are several things that could go wrong here. The drain line could get _____, causing water to leak out. The drain pan could get clogged, causing it to overflow. Or pump could stop working. Any of those things could cause water to end up in your duct, and eventually drip out _____ your AC vent.

| dependent | liquify | clog | empty | drain |
| accumulate | empty | That | by-product | through |

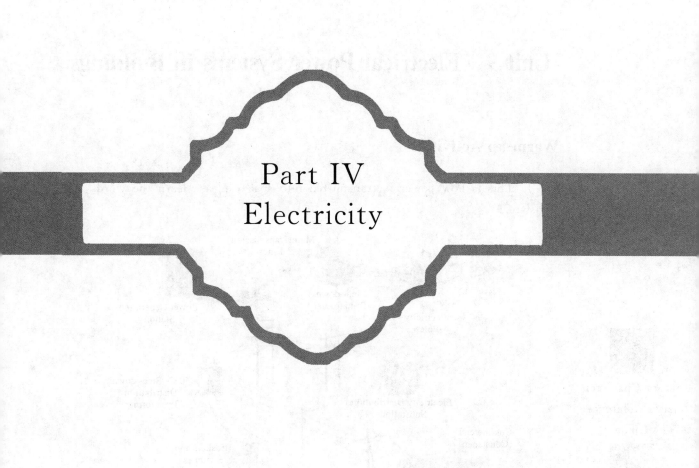

Part IV
Electricity

Unit 9   Electrical Power Systems in Buildings

Warm-up Activities

1. This is how power is transmitted into space, please learn the words and translate.

Sensor Characteristics Reference Guide

Method and Techniques of Electricity

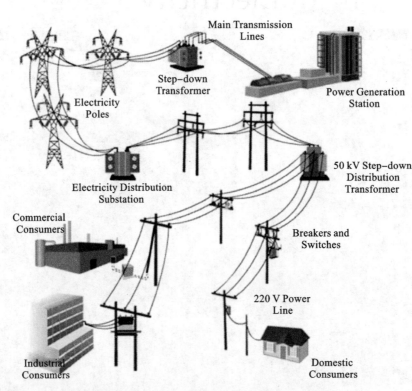

(Source: Hussain Z., Me-mon S., Shah, R., et al)

2. Fill in the blanks with the words or expressions given below, and change the form where necessary.

In a power _____ system, these principal components must be contained, _____ generating station, substations, transmission _____ and cable, and industrial, residential loads. The substations have different functions. The interconnecting substations serve to _____ different power systems together. The distribution substations change the medium voltage to low voltage to change the line voltage _____ step-down transformers. The transmission substations _____ to change the line voltage and regulate it. These substations also _____ some protective circuits to _____ quick isolation of faulted lines from the system. The lines can be classified into four types _____ their voltage class. They are low-voltage lines, _____ lines, high-voltage lines and extra-high-voltage lines.

| such as | medium-voltage | contain | distribute | serve |
| tie | by means of | according | provide | lines |

3. Translate Chinese into English.

(1) 电能通过架空线和地下电缆传输。

(2) 供配电系统必须提供变化幅度不超过额定电压±10%的电压。

(3) 据电压等级通常把电力线分为四类。

Intensive Reading

History of Electric Power

Benjamin Franklin is known for his discovery of electricity. Born in 1706, he began studying electricity in the early 1750s. His observations, including his kite experiment, verified the nature of electricity. He knew that lightning was very powerful and dangerous. The famous 1752 kite experiment featured a pointed metal piece on the top of the kite and a metal key at the base end of the kite string. The string went through the key and attached to a Leyden jar. He held the string with a short section of dry silk as insulation from the lightning energy. He then flew the kite in a thunderstorm. He first noticed that some loose strands of the hemp string stood erect, avoiding one another. He proceeded to touch the key with his knuckle and received a small electrical shock.

Leyden jar
莱顿瓶

Between 1750 and 1850 there were many great discoveries in the principles of electricity and magnetism by Volta, Coulomb, Gauss, Henry, Faraday, and others. It was found that electric current produced a magnetic field and that a moving magnetic field produced electricity in a wire. This led to many inventions such as the battery, generator, electric motor, telegraph, and telephone, plus many other intriguing inventions. In 1879, Thomas Edison invented a more efficient lightbulb, similar to those in use today. In 1882, he placed into operation the historic Pearl Street steam—electric plant and the first direct current (DC) distribution system in New York City, powering over 10,000 electric lightbulbs. By the late 1880s, power demand for electric motors required 24-hour service and dramatically raised electricity demand for transportation and other industry needs.

By the end of the 1880s, small, centralized areas of electrical power distribution were sprinkled across U.S. cities [1]. Each distribution center was limited to a service range of a few blocks because of the inefficiencies of transmitting direct current. Voltage could not be increased or decreased using direct current systems, and a way to transport power over longer distances was needed. To solve the problem of transporting electrical power over long distances, George Westinghouse developed a device called the "transformer". The transformer allowed electrical energy to be transported over long distances efficiently. This made it possible to supply electric power to homes and businesses located far from the electric generating plants. The application of transformers required the distribution system to be of the alternating current (AC) type as opposed to direct current (DC) type. The development of the Niagara Falls hydroelectric power plant in 1896 initiated the practice of placing electric power generating plants far from consumption areas. Niagara was the first large power system to supply multiple large consumers with only one power line.

Since the early 1900s alternating current power systems began appearing throughout the United States. These power systems became interconnected to form what we know today as the three major power grids in the United States and Canada. The following are the fundamental terms used in today's electric power systems based on this history.

System Overview

Electric power systems are real-time energy delivery systems. Real time means that power is generated, transported, and supplied the moment you turn on the light switch [2]. Electric power systems are not storage systems like water systems and gas systems. Instead, generators produce the energy as the demand calls for it.

Fig. 9-1 shows the basic building blocks of an electric power system. The system starts with generation, by which electrical energy is produced in the power plant and then transformed in the power station to high-voltage electrical energy that is more suitable for efficient long-distance transportation. The power plants transform other sources of energy in the process of producing electrical energy. For example, heat, mechanical, hydraulic, chemical, solar, wind, geothermal, nuclear, and other energy sources are used in the production of electrical energy. High-voltage (HV) power lines in the transmission portion of the electric power system efficiently transport electrical energy over long distances to the consumption locations. Finally, substations transform this HV electrical energy into lower-voltage energy that is transmitted over distribution power lines that are more suitable for the distribution of electrical energy to its destination, where it is again transformed for residential, commercial, and industrial consumption.

Electric utilities transmit power from the power plant most efficiently at very high voltages. In the United States, power companies provide electricity to medium or large buildings at 13,800 volts (13.8 kV). For small commercial buildings or residential customers, power companies lower the voltage with a transformer on a power pole or mounted on the ground. From there, the electricity is fed through a meter and into the building.

electric utilities
电力设施

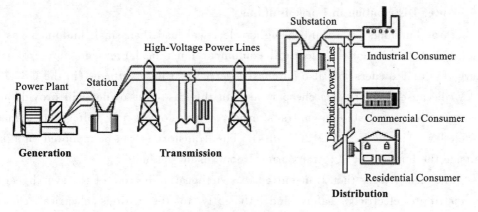

Figure 9-1 System overview

Transmission of Electrical Energy

Power Distribution in Small Buildings

Small commercial or residential buildings have a very simple power distribution system (as in Fig. 9-2). The utility will own the transformer, which will sit on a pad outside the building or will be attached to a utility pole. The transformer reduces the voltage from 13.8 kV down to 120/240 or 120/208 V and then passes the electricity to a meter, which is owned by the utility and keeps a record of power consumption.

After leaving the meter, the power is transmitted into the building at which point all wiring, panels, and devices are the property of the building owner. Wires transfer the electricity from the meter to a panel board, which is generally located in the basement or garage of a house. In small commercial buildings, the panel may be located in a utility closet. The panel board will have a main service breaker and a

transformer
变压器
utility pole
电线杆
meter
仪表

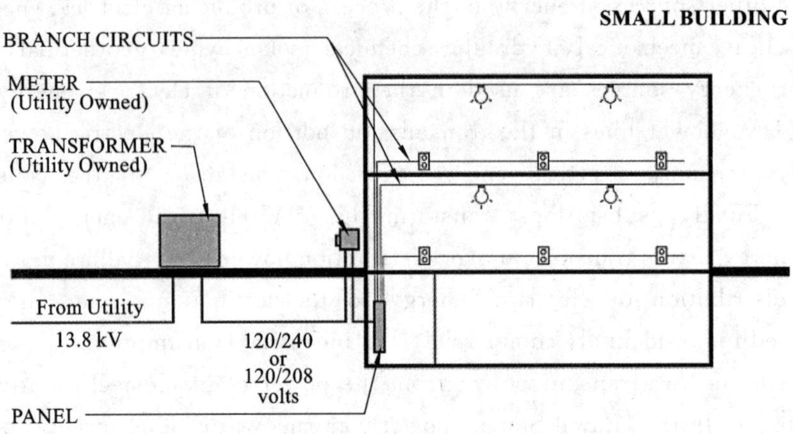

```
                                                        SMALL BUILDING
                  BRANCH CIRCUITS
                  METER
                  (Utility Owned)
                  TRANSFORMER
                  (Utility Owned)

         From Utility
         13.8 kV      120/240
                        or
                      120/208
                       volts
                PANEL
```

Figure 9-2 Power distribution in small buildings

series of circuit breakers, which control the flow of power to various circuits in the building [3]. Each branch circuit will serve a device (some appliances require heavy loads) or a number of devices like convenience outlets or lights.

Power Distribution in Large Buildings

Large buildings have a much higher electrical load than small buildings (as in Fig. 9-3); therefore, the electrical equipment must be larger and more robust. Large building owners will also purchase electricity at high voltages (in the US, 13.8 kV) because it comes at a cheaper rate. In this case, the owner will provide and maintain their own step-down transformer, which lowers the voltage to a more usable level (in the US, 480/277 volts). This transformer can be mounted on a pad outside the building or in a transformer room inside the building.

The electricity is then transmitted to switchgear. The role of the switchgear is to distribute electricity safely and efficiently to the various electrical closets throughout the building [4]. The equipment has numerous safety features including circuit breakers, which allow power to be disrupted downstream — this may occur due to a fault or problem, but it can also be done intentionally to allow technicians to work on specific branches of the power system.

It should be noted that very large buildings or buildings with complex electrical systems may have multiple transformers, which may feed multiple pieces of switchgear.

The electricity will leave the switchgear and travel along a primary feeder or bus. The bus or feeder is a heavy gauge conductor that is capable of carrying high amperage current throughout a building safely and efficiently. The bus or feeder is tapped as needed and a conductor is run to an electric closet, which serves a zone or floor of a building.

---

convenience outlets
便利店

switchgear
开关设备

circuit breaker
断路器

primary feeder
一次给料机

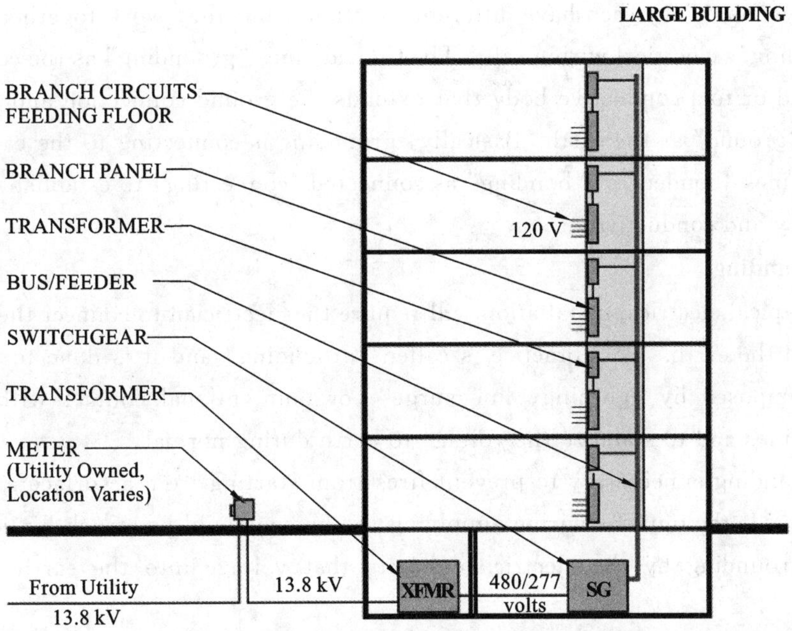

Figure 9-3　Power distribution in large buildings

　　Each electrical closet will have another step-down transformer — in the US, this will drop the power from 480/277 volts to 120 volts for convenience outlets. That transformer will feed a branch panel, which controls a series of branch circuits that cover a portion of the building. Each branch circuit covers a subset of the electrical needs of the area — for instance, lighting, convenience outlets to a series of rooms, or electricity to a piece of equipment[5].

electrical closet
电气柜

Notes
[1]到 19 世纪 80 年代末,美国各城市集中电力分布尚不普及。

[2]电力系统为实时能量传输系统。实时意味着你一打开电灯开关,电力就产生、输送和供应。

[3]配电盘将有一个主服务断路器和用于控制建筑物内各电路功率的一系列断路器。

[4]配电盘的作用是将电力安全有效地分配到整个建筑物的各个电气柜。

[5]每个分支电路负责该区域的某一个子设备的电气需求,例如,照明、插座或设备供电。

Extensive Reading

Grounding & Bonding

　　"Grounding" and "bonding" are important elements of a building's electrical

Grounding

wiring system. They each have different functions, but they work together to make the building's electrical wiring safe. The Code defines "grounding" as the connecting to ground or to a conductive body that extends the ground connection and the Code defines "ground" as the earth. Basically, grounding is connecting to the earth. The Code defines "bonded" or "bonding" as connected (connecting) to establish electrical continuity and conductivity.

Grounding

A typical electrical installation will require the electrician to connect the house's wiring to the earth. This practice is called "grounding" and it is done to limit the voltage imposed by lightning, line surges, or unintentional contact with higher-voltage lines and to stabilize the voltage to earth during normal.

Grounding is necessary to prevent fires from starting from a surface arc within the home. If the outdoor wiring supplying the home should be struck by lightning, proper grounding by the electrician directs that voltage into the earth where it dissipates.

Because the Code requires the neutral conductor of a single-phase, 3-wire system to be grounded, in our example, the electrician will normally use a bare copper wire to connect the neutral conductor (often called the grounded conductor) to a grounding electrode that has direct contact with the earth.

Each premises wiring system supplied by a grounded AC service shall have a grounding electrode conductor connected to the grounded service conductor at each service. Basically, this is saying each building that is served electricity shall have its service connected to the earth.

The Code allows this grounding connection (earth connection) to be at various locations. It says the grounding electrode conductor connection shall be made at any accessible point from the load end of the service drop or service lateral to the terminal or bus to which the grounded service conductor is connected at the service disconnecting means.

Please note that the Code does not indicate that a meter base enclosure is not an accessible location for terminating a grounding electrode conductor. These enclosures are sometimes viewed this way because of the seal that the cooperative puts on them to place a legal jurisdiction over the unmetered conductors contained in the enclosure. However, any way you look at this connection location, it is still accessible (by definition) to workers and others, by legal means. In other words, just notify the cooperative and access can be granted beyond the seal. Besides, this earthing connection is made in the meter base before it is energized and before the seal or lock is installed. Some local jurisdictions may not permit the connection to be made in the meter base enclosure, but that is not the intent of the Code — the Code does allow it.

From Grounding to Bonding

Grounding and bonding work together with each other to make a building's electrical wiring safe. Once the electrician has completed the connection of the grounded conductor to the earth, the focus now moves from the concept of grounding to bonding. This transition is started through use of a device called the Main Bonding Jumper.

250.24(B) of the Code requires the electrician to connect the grounded service conductor to the metal enclosure of the Service Equipment and to the equipment grounding terminal bar within the Service Equipment. This connection can be accomplished through various types of Main Bonding Jumpers, including a wire, bus, screw, or similar suitable conductor.

Once grounding has been performed and the Main Bonding Jumper has been connected, the grounded conductor (neutral) which is attached to the neutral terminal bar in the Service Equipment is also now connected to the earth metal enclosure of the Service Equipment, grounding electrode conductor, and equipment grounding terminal bar of the Service Equipment.

Bonding

Once the electrician has completed grounding of the Service Equipment and the attachment of the Main Bonding Jumper within the Service Equipment, it's time to perform another critical step in the house's electrical wiring. As the electrician installs the branch circuits in the house, focus is now concentrated on making sure that the metal components in all of the branch circuits are connected to the metal enclosure of the Service Equipment — an act called "bonding". Bonding is accomplished when the electrician connects equipment grounding conductors to the equipment grounding terminal bar in the Service Equipment and to the grounding terminals on electrical devices and electrical boxes. If the electrician is using metal raceways and metal boxes to install conductors, the electrician can also use the metal raceway conduit as the equipment grounding conductor, but great care needs to be taken when using this method. Connections at each conduit joint and box must use proper equipment and methods to ensure a solid connection has been made. When installed correctly, here is a way to visualize what proper bonding does. Imagine taking one lead of a continuity tester and touch a metal component of one circuit (box, enclosure, conduit, locknut, appliance frame, etc.), and then take the other (extremely long!) lead anywhere in the building and touch a similar metal component used either in the same circuit or in a different electrical circuit. If bonding was properly performed, the continuity tester would show continuity between the two places being touched. Basically, all metal components have become "one unit". That is why even when one is using plastic switch boxes and plastic switch covers for installing switches, the electrician is required to use switches with a grounding terminal and conductors that include an equipment grounding conductor. Because of the possibility of a metal switch cover being used some day, the electrician once again has to ensure that the metal cover has a conductive path back to all metallic parts of its circuit and other circuits and the metal enclosure of the Service Equipment. You may be wondering if one is using equipment "grounding" conductors and connecting to "grounding" terminals, then why isn't this considered an act of grounding? It's a great question. The best way to answer it is to say it's really both bonding and grounding. Grounding and bonding happen simultaneously during the installation of an electrical system and branch circuits. They work hand in hand in order to make it safe.

▶▶▶ Post-Reading Exercises

1. Fill in the blanks with the words or expressions given below, and change the form where necessary.

At places _____ the grounding resistance obtained with this arrangement exceeds the _____ limit, multiple electrodes should be used. In case of two electrodes the interconnection should be made with a MS strip of the same size _____ the grounding conductor and the distance _____ them should not be less than twice the length of the electrode. If the use of a third electrode is also necessitated, it should be placed _____ the three electrodes when interconnected form an equilateral triangle with sides not less than twice the length of the electrode. The _____ two or three rods can be considered _____ parallel for practical purposes and the total earth resistance will thus become half or _____ third of a single rod earth resistance respectively. Sometimes, it may be difficult to keep such large spacing and so it becomes necessary _____ determine what reduction in the total resistance can be obtained _____ connecting rods in parallel.

| so that | up | one | by | to |
| as | where | between | prescribe | in |

2. Translate Chinese into English.

(1) 电气设备保护方案中的一个关键因素是接地。

(2) 接地极的长度增加一倍,电阻就减少 40%,而直径增加一倍,电阻的减少量小于 10%。

(3) 所有的表达式都假定土壤中的电阻率一致。

(4) 如果土壤中含有不溶性盐,则潮湿的土壤也可能具有较高的电阻率。

(5) 在一些地方,获得 5Ω 甚至更小的电阻不太困难,而另一些地方获得 100Ω 的电阻相当困难。

Fire Alarm Circuit

Fire alarm circuit, as its name implies, sounds an alarm in the event of fire. There can be one or several alarms throughout a building, and there can be several alarm points, which activate the warning. The alarm points can be operated manually or automatically; in the latter case they may be sensitive to heat, smoke or ionization. There are clearly many combinations possible, and this paper gives some systematic account of the way they are built up.

The simplest scheme is shown in Fig. 9-4. Several alarm points are connected in parallel, and whenever one of them is actuated the circuit is completed and the alarm sounds. This is described as an open circuit, and it will be seen that it is not fail safe, because if there is a failure of supply the fire alarm cannot work. Another characteristic of this circuit is that every alarm point must be capable of carrying the full current taken by all the bells or hooters working together.

A slightly more elaborate scheme is shown in Fig. 9-5. The alarm points are connected in series with each other and with a relay coil. The relay is normally closed when de-energized, and opens when the coil is energized. Thus when an alarm point is activated the relay coil is de-energized, the relay closes and the alarm sounds. This system fails safe to the extent that if the coil circuit fails the main circuit operates the alarm. It is not of course safe against total failure of the supply because in that event there is no supply available to work the bells. The alarm points do not have to carry the operating current of the bells or hooters. This arrangement is called a closed circuit in contrast to the open circuit of Fig. 9-4. We can notice that in an open circuit the alarm points are wired in parallel and are normally open, whilst in a closed circuit they are wired in series and are normally closed.

A typical manually operated fire alarm point is contained in a robust red plastic case with a glass cover. The material is chosen for its fire resisting properties. The case has knock out for conduit entries at top and bottom but the material can be sufficiently easily cut for the site electrician to make himself an entry in the back if he needs it. Alternative terminals are provided for circuits in which the contacts have to close when the glass is smashed (as in Fig. 9-4) and for circuits in which the contacts have to open when the glass is smashed (as in Fig. 9-5). In the former case, there is a test switch which can be reached when the whole front is opened with an Allen Key. In the latter case, the test push is omitted because the circuit is in any case of the fail safe type.

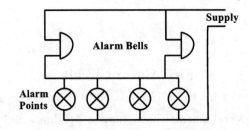

Figure 9-4   Fire alarm open circuit

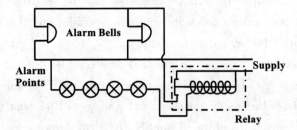

Figure 9-5   Fire alarm closed circuit

The alarm point illustrated is suitable for surface mounting. Similar ones are available for flush fixing and in weatherproof versions. The current carrying capacity of the contacts should always be checked with the maker' catalogue.

A thermally operated alarm point consists of a bi-metal strip that deflects when the temperature rises, and thereby tilts a tube half full of mercury. When the tube is tilted the mercury flows into the other half of the tube where it completes the circuit between two contacts previously separated by air. Alternatively, the arrangement within the tube can be such that the mercury breaks the circuit when the tube is tilted. The casing of the alarm is of stainless steel. Heat detectors of this type are usually set to operate at 65℃. They are frequently used in boiler houses.

A smoke operated alarm point would be used only in special circumstances which make it necessary to detect smoke rather than heat. This type can cause nuisance operation of the alarm by reacting to small quantities of smoke which have not been caused by a fire; they have for example been known to sound the alarm as a result of cigarette smoke in an office. Modern ones have adjustable sensitivity so that they can be set to avoid nuisance operation.

An ionization detector contains a chamber which houses some low strength radioactive material and a pair of electrodes. The radioactive material makes the air in the chamber conductive so that a small current flows between the electrodes. The size of the current varies with the nature of the gas in the chamber and as soon as any combustion products are added to the air there is a sudden change in the current flowing. The detector also has a second chamber which is permanently sealed so that the current through it never changes. As long as the currents through the two chambers are equal there is no output, as soon as they become unbalanced there is a net output which is used to operate a transistor switch in the main circuit through the detector.

》》》 **Post-Reading Exercises**

1. Choose the best answer for each of the following.
   (1) Several alarm points are connected in _____ and this is described as an open circuit.
      A. parallel
      B. series
      C. parallel and series
   (2) In this paper, the author introduces _____ kinds of operated alarm point.
      A. 2
      B. 3
      C. 4
   (3) An ionization detector also has a second chamber which is permanently sealed so that the current through it _____.
      A. reduces
      B. increases
      C. never changes
   (4) When an alarm point is activated the relay coil is _____, the relay _____.
      A. de-energized; opens
      B. de-energized; closes
      C. energized; closes

(5) When it is necessary to detect smoke rather than heat, the system should be used _____.
   A. thermally operated alarm point
   B. smoke operated alarm point
   C. an ionization detector

2. Translate Chinese into English.

(1) 报警点以相互串联的形式连接在一起,并串接了一个中继器。

(2) 材料的选择取决于它的防火性能。

(3) 该型号感温探测器的设置的温度为65℃。

(4) 现今该探测器已经能够调整灵敏度,从而可以避免它的误动作。

(5) 离子探测器包括两部分:一个由低放射性材料制成的电离室和一对电极。

3. Fill in the blanks with the words or expressions given below, and change the form where necessary.

Electrically operated sirens can be used _____ an alternative to bells. They are _____ louder and can be heard at a distance of half _____ three quarters of a mile. This makes them very suitable _____ factories in which there may be a lot of background noise. A typical _____ is 60 W and we see that a siren takes a much bigger current _____ a bell. As a consequence, there is a bigger voltage _____ in the circuit feeding it. In low voltage circuits, the drop can be sufficiently _____ to prevent the sirens _____ sounding at all, and it becomes especially important to check _____ for voltage drop.

| to | rating | from | as | layout |
| drop | serious | much | for | than |

Unit 10 Building Automation System

Warm-up Activities

1. Learn the words in the diagram.

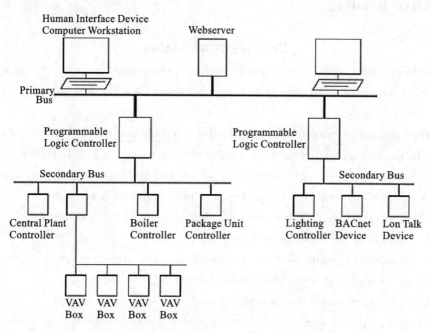

(Source: Cree, J. V., Dansu, A., Fuhr, P. et al)

2. Fill in the blanks with the words or expressions given below, and change the form where necessary.

The intelligence of the early monitoring systems _____ centralized at the primary central processing unit with the sensing and commands _____ switch plant via remote non-intelligent data gathering panels located at the vicinity of the plant or process being monitored and controlled. These panels incorporated their _____ sensing devices and as with the electro-mechanical systems, they oversaw the control systems operating the plant.

Transmission _____ data between the central processor and data gathering panels was multiplexed over a twisted pair _____ in a digital format allowing for _____ transfer of data and the ability to monitor and control a considerable number of devices and elements. Their primary purpose was to _____ the information _____ alarms, status of plant and variables, such as temperature and humidity and electrical consumption. Various ranges of inputs they provided the means to centralize the automatic and manual control of the plant against variables _____ time, outdoor temperature and energy saving programs.

| was | cable | of | fast | from |
| such as | own | centralize | in respect of | to |

Intensive Reading

Building Automation

Building automation most broadly refers to creating centralized, networked systems of hardware and software monitors and controls a building's facility systems (electricity, lighting, plumbing, HVAC, water supply, etc.) [1].

When facilities are monitored and controlled in a seamless fashion, this creates a much more reliable working environment for the building's tenants [2]. Furthermore, the efficiency introduced through automation allows the building's facility management team to adopt more sustainable practices and reduce energy costs.

These are the four core functions of a building automation system:
1. To control the building environment.
2. To operate systems according to occupancy and energy demand.
3. To monitor and correct system performance.
4. To alert or sound alarms when needed.

At optimal performance levels, an automated building is greener and more user-friendly than a non-controlled building.

What Is Meant By "Controlled"?

A key component in a building automation system is called a controller, which is a small, specialized computer. Controllers regulate the performance of various facilities within the building. Traditionally, this includes the following:
- Mechanical systems.
- Electrical systems.
- Plumbing systems.
- Heating, ventilation and air-conditioning systems.
- Lighting systems.
- Security & surveillance systems.

---

building automation 楼宇自动化

security systems 安全系统
surveillance systems 监视系统

A more robust building automation system can even control security systems, the fire alarm system and the building's elevators.

To understand the importance of control, it helps to imagine a much older system, such as an old heating system. Take wood-burning stoves, for example. Anyone heating their buildings through pure woodfire had no way to precisely regulate the temperature, or even the smoke output. Furthermore, fuelling that fire was a manual effort. Fast-forward 150 years: Heating systems can be regulated with intelligent controllers that can set the temperature of a specific room to a precise degree [3]. And it can be set to automatically cool down overnight, when no one is in the building.

The technology that exists today allows buildings to essentially learn from itself. A modern building automation system will monitor the various facilities it controls to understand how to optimize for maximum efficiency. It is no longer a matter of heating a room to a specific temperature; systems today can learn who enters what rooms at what times so that buildings can adjust to the needs of the tenants, and then conserve energy when none is needed.

There is a growing overlap between the idea of controlling a building and learning from all the data the system collects. That is why automated buildings are called "smart buildings" or "intelligent buildings". They're getting smarter all the time.

The Evolution of Smart Buildings

Kevin Callahan, writing for Automation.com, points to the creation of the incubator thermostat — to keep chicken eggs warm and allow them to hatch — as the origin of smart buildings.

Like most technologies, building automation has advanced just within our lifetimes at a rate that would have baffled facility managers and engineers in, say, the 1950s. Back then, automated buildings relied on pneumatic controls in which compressed air was the medium of exchange for the monitors and controllers in the system.

By the 1980s, microprocessors had become small enough and sufficiently inexpensive that they could be implemented in building automation systems. Moving from compressed air to analog controls to digital controls was nothing short of a revolution. A decade later, open protocols were introduced that allowed the controlled facilities to actually communicate with one another. By the turn of the millennium, wireless technology allowed components to communicate without cable attachments.

How Do Building Automation Systems Work?

Basic BAS has five essential components:

• **Sensors** —Devices that measure values such as CO_2 output, temperature, humidity, daylight or even room occupancy.

• **Controllers** — These are the brains of the systems. Controllers take data from the collectors and decide how the system will respond.

• **Output devices** — These carry out the commands from the controller. Example devices are relays and actuators.

• **Communications protocols** — Think of these as the language spoken among the components of the BAS. A popular example of a communications protocol is BAC net.

• **Dashboard or user interface** — These are the screens or interfaces humans use to interact with the BAS. The dashboard is where building data are reported.

The Role of Controllers

Controllers are the brains of the BAS, so they require a little more exploration. As mentioned above, the advent of direct digital control modules opened up a whole universe of possibilities for automating buildings. A digital controller can receive input data, apply logic (an algorithm, just as Google does with search data) to that information, then send out a command based on what information was processed [4]. This is best illustrated through the basic three-part DDC loop:

1. Assume a sensor detects an increase in temperature in a company's board room when the room is known to be unoccupied.

2. The controller will apply logic according to what it knows: no one is expected in that room, thus there is no demand for additional heat, thus there is no need for that room to warm up. (Note: The algorithm with which a controller processes information is actually far more complex than depicted in this example.) It then sends a command to the heating system to reduce output.

3. The actual heating unit for the boardroom in question receives that command and dials back its heat output. All of these appear to happen almost instantaneously.

Why Are Building Automation Systems Useful?

The benefits of building automation are manifold, but the real reasons facility managers adopt building automation systems break down into three broad categories:

1. They save building owners money.

2. They allow building occupants to feel more comfortable and be more productive.

3. They reduce a building's environmental impact.

Saving Money

The place where a BAS can save a building owner a significant amount of money is in utility bills. A more energy-efficient building simply costs less to run. Building and room occupancy, as demonstrated earlier with the heated boardroom example. If a building can know when the demand for lighting or HVAC facilities will wax and wane, then it can dial back output when demand is lower. Estimated energy savings from simply monitoring occupancy range from 10-30%, which can add up to thousands of dollars saved on utilities each month.

Furthermore, a building can also sync up with the outdoor environment for maximum efficiency. This is most useful during the spring and summer, when there is more daylight (and thus less demand for interior lighting) and when it is warmer outside, allowing the building to leverage natural air circulation for comfort.

sync up
同步

Data collection and reporting also makes facility management more cost efficient. In the event of a failure somewhere within the system, this will get reported right on the BAS dashboard, meaning a facility professional does not have to spend time looking for and trying to diagnose the problem.

data collection
数据收集

Finally, optimizing the operations of different building facilities extends the lives of the actual equipment, meaning reduced replacement and maintenance costs.

Comfort and Productivity

Smarter control over the building's internal environment will keep occupants happier, thereby reducing complaints and time spent resolving those complaints. Furthermore, studies have shown that improved ventilation and air quality have a direct impact on a business's bottom line: Employees take fewer sick days, and greater comfort allows employees to focus on their work, allowing them to increase their individual productivity [6].

Environmentally Friendly

The key to an automated building's reduced environmental impact is its energy efficiency. By reducing energy consumption, a BAS can reduce the output of greenhouse gases and improve the building's indoor air quality, the latter of which ties back into bottom-line concerns about occupant productivity. Furthermore, an automated building can monitor and thus control waste in facilities such as the plumbing and wastewater systems. By reducing waste through efficiencies, a BAS can leave an even smaller environmental footprint. In addition, a regulatory government agency could collect the BAS's data to actually validate a building's energy consumption. This is key if the building's owner is trying to achieve LEED or some other type of certification.

Notes

[1]从广义上说,建筑自动化是指通过软硬件的集中网络化系统来监控和控制建筑的设施系统(电力、照明、管道、暖通空调、供水等)。

[2]当设施以无缝的方式进行监控和控制时,这就为大楼的租户创造了一个更加

可靠的工作环境。

［3］快进 150 年：供暖系统可以用智能控制器调节，可以精确设定特定房间的温度。

［4］数字控制器可以接收输入数据，对该信息应用逻辑（一种算法，就像谷歌搜索），然后根据处理的信息发出命令。

［5］楼宇自动化的好处是多方面的，但设备经理采用楼宇自动化系统的真正原因有三大类。

［6］研究表明，通风和空气质量的改善直接影响到企业的效益：员工请的病假更少，提供更舒适的环境使员工能够专注于自己的工作，从而提高个人生产力。

Extensive Reading

Intellegence Building

Intelligent Buildings

Today, almost every new building has to accommodate higher levels of servicing than ever before. Although some buildings completed recently have been labelled intelligent, there is still a fair way to go before integrated "smart" are likely to be capable of maintaining optimal internal environments. However, this issue is just one of many that must be considered when attempting to characterize intelligent buildings.

Several different types of technology need to be present before any building could be termed intelligent. They exist in many buildings already, although have often been introduced as after-thoughts, instead of resulting from the identified needs of owners and occupants. However, many of the problems are not related to the technology—this provides the solutions—rather, they are related to the needs of owners and occupants.

A recent report on the Japanese construction industry identified three attributes that an intelligent building should possess.

• Buildings should "know" what is happening inside and immediately outside.

• Buildings should "decide" the most effective way of providing a convenient, comfortable and productive environment for the occupants.

• Buildings should "respond" quickly to occupants requests.

These attributes may be translated into a need for various technology and management systems. The successful integration of these systems will produce the intelligent buildings, containing:

• Building automation (BA) systems: to enable the building to respond to external factors and conditions (not just climatic, but also fire and security protection); simultaneous sensing, control and monitoring of the internal environment; and the storage of the data generated, as knowledge of the building's performance, in a central computer system.

• Office automation (OA) systems: to provide management information and link to the central computer system as decision support aids.

• Communications automation (CA) systems: to enable rapid communication with the outside world, via the central computer system, using optical fibre installations, microwave and conventional satellite links.

A consequence of this integration is that intelligent buildings will have more in common with engineering projects than those of traditional construction. They require many different skills and an understanding of technology in broader context than previously needed has hitherto been the need within the construction industry.

The notion of the intelligent building is linked increasingly with big business, where being a part of a world market demands considerable inter-organization communications and a building that can deliver them. A culture of high salaries and ultra-dependence upon extensive, expensive information technology has become common after the Big Bang in the UK.

The potential obsolescence inherent in anything containing a computer means that operating costs for those buildings can be many times greater than those of just few years ago. Internal environments have to be controlled more precisely; the penalty for any failure in communications, for instance, can be catastrophic. Furthermore, the capital cost of some of these buildings could conceivably be less than one day's trading on the international market. This is the real backdrop for many potential intelligent building projects. They are demanded by clients who can probably afford better buildings than the industry is currently able to offer.

A future trend evident from a study of the Japanese industry is the extent of prefabrication and industrialized buildings; an approach that evokes many unhappy memories in the UK. Bad experiences of industrialized building systems still linger and are not helped by periodic reminders occasioned by the demolition of yet another local authority tower block.

But industrialized systems are different today. One important distinction is that yesterday's industrialized buildings were concerned mainly with the prefabrication of entire building superstructures, incorporating relatively primitive services installations. Nowadays prefabrication is likely to mean provision of complete, highly serviced areas within the building and systems for retaining flexibility in the layout of internal spaces.

A trend towards a component-based approach to design is also evident, encouraged to an increasing extent by computer aided design (CAD) techniques. By using CAD systems, it is possible to compose designs from standard components held in an on-line library. As a result, drafting time and cost can be reduced significantly on some projects.

A further distinction today is that superstructure can be constructed rapidly using means other than the prefabrication of major components. The innovative procurement methods and designs that deliberately take account of the construction process are just two examples, and are often used in combination.

> **Post-Reading Exercises**

1. Choose the best answer for each of the following.

(1) An intelligent building should can _____.

    A. do everything by itself

    B. know what is happening inside and outside

    C. decide environment changes

(2) In this paper, the intelligent building is defined as _____ systems.

    A. 3A

    B. 4A

    C. 5A

(3) _____ is not the function of BA.

    A. Simultaneous sensing

    B. Monitoring the internal environment

    C. Provide management information

(4) By using CAD systems, drafting time is _____ and cost is _____ on some projects.

    A. increased; reduced

    B. reduced; increased

    C. reduced; reduced

(5) _____ is not the transport medium in communication automation system.

    A. Coaxial line

    B. Optical fibre

    C. Microwave

2. Translate Chinese into English.

(1) 智能建筑为人们提供一个方便、舒适和高效环境。

(2) 智能建筑的发展和先进计算机技术密切相关。

(3) 办公自动化系统常与专业计算机系统相连,为做出正确决策提供管理信息。

(4) 通信自动化系统能够使建筑物与外界进行快速通信。

(5) 楼宇自动化系统应对建筑内部环境进行同步的传感、控制和监视。

Smart Control of HVAC System

It takes an enormous amount of energy to condition air and then distribute it throughout a building; not surprisingly, HVAC equipment typically consumes at least 40% of a commercial building's energy. Many buildings' HVAC systems consume even more energy than that, as roughly one-third of them are oversized for the space they serve. Using controls to properly manage HVAC operation is an essential part of saving energy in a building. However, building operators frequently manage HVAC operations through trial-and-error adjustments in reaction to occupant comfort feedback—sometimes relegating energy savings to a much lower priority. Smart HVAC systems have the potential to greatly reduce energy consumption while maintaining or even improving occupant comfort. Smart building software interprets information from a variety of HVAC sensor points and maintains that information in real time, in a cloud-based system that is remotely accessible. Engineers develop algorithms within the smart building software that use the database information to optimize the monitoring and control of HVAC systems. These advanced controls can limit HVAC consumption in unoccupied building zones, detect and diagnose faults, and reduce HVAC usage during times of peak energy demand.

Using Sensors to Optimize Operations

Sensors are devices that sense a physical stimulus and convert it into a signal. One example of a sensor that provides HVAC systems with useful data is a duct static pressure sensor, which measures the amount of resistance against the air flowing through a duct. HVAC fans must work harder to overcome greater resistance in ducts, so reducing static pressure saves energy. A duct static pressure sensor contains a sensing element that reacts to physical changes—in this case, static pressure. The sensor transmits an electrical signal that indicates a change in duct static pressure. Building operators can use these static pressure readings to control the HVAC systems to operate at a particular duct static pressure, and can even use the measurement to identify HVAC system faults. The falling cost of sensor technologies is a key catalyst of smart HVAC. In 2004, a basic Internet of Things (IoT) sensor—such as your phone's little gyroscope sensor—cost \$1.30 on average; by 2014, a basic sensor's cost averaged less than \$0.60. With lower technology costs and the increasing availability of wireless technologies, it is now easier than ever to cheaply obtain sensor readings for various HVAC components. Further, recent advancements in data storage and cloud computing make it possible for building operators to access the multitude of HVAC data points, such as temperature, pressure, flow rate, and gas concentration. One of the largest energy efficiency benefits of smart building HVAC controls is found through optimizing the amount of conditioned (i.e., heated or cooled) air supplied throughout a building. Although it may seem like a simple concept, this goal can be achieved in various ways. For example, a whole-building ventilation controls system, with smart capability, senses the amount of carbon dioxide (CO_2) in occupied areas of the building and can modulate the amount of airflow in one area without starving or over-ventilating another. This can save considerable energy in heating and cooling and ventilation fan operation. In addition to controlling HVAC operation based on CO_2 levels, smart controls can optimize air flow using data provided by occupancy, temperature, humidity, duct static pressure, and air quality sensors. Small-and medium-sized buildings may not have the funding to invest in whole-building HVAC control systems and may be more inclined to install smart controls directly on HVAC equipment. For example, the Pacific Northwest National Laboratory evaluated the effectiveness of rooftop units (RTUs) retrofitted with advanced controllers equipped with multi-speed fan, economizer, and demand ventilation controls. The study found approximately 50% electricity savings for RTUs with a three-year payback, even in regions with low electricity prices. This smart RTU technology was deemed so successful that is was subsequently included in California's Title 24 and DOE's RTU manufacturing standard. Smart HVAC systems can also support sophisticated data analysis. Historically, building operators of a typical commercial building have been limited to reviewing rudimentary energy bill data. This form of data analysis is limited because the operator has reduced visibility into actual systems performance and interactions, often relying on month-old whole-building meter data. Armed with smart building data analytics, building operators can review historical building occupancy and usage on a granular level, receive performance data in real time and fine-tune the HVAC controls accordingly, thereby avoiding wasted HVAC usage.

Controlling Multiple Zones

Optimizing the use of conditioned air is one of the most effective applications of smart building equipment, especially in multi-zone systems. For example, a multi-zone variable air volume (VAV) system with six VAV boxes could use smart controls to more effectively condition each of the six zones. With sensors installed in each office area, each VAV box can be programmed to cycle back or shut off completely when the corresponding space is vacant. If most of the employees are out of the office, the smart controls can reduce or shut off conditioning to any or all of the six zones to save energy. An example of a less efficient alternative is a whole floor with a constant air volume (CAV) system served by one air-handling unit. In this case, control options are limited to cycling back or turning off the airflow on the entire floor, and the flexibility of zone control is lacking.

Hotel rooms, which could be considered individual zones, offer a unique application for zone control energy savings. Hotel room HVAC units typically remain set at whatever temperature the guest or cleaning staff last selected, thus conditioning unoccupied guest rooms. This can waste hundreds of thousands of kWh of electricity and tens of thousands of dollars in energy costs each year. Connecting guest room vacancy controls and HVAC system controls allows hotels to set back the temperature when the room is vacant. Of course, building operators must balance these energy-saving measures with occupant comfort to ensure that guests do not experience discomfort when they return to their rooms. Navigant estimated that only about 30% of the global hospitality industry currently uses room-based energy management systems to reduce HVAC consumption. This figure is likely even lower in the United States, because relatively few US hotels contain smart HVAC controls compared to hotels in many parts of Asia and Europe.

⟫⟫⟫ Post-Reading Exercises

1. Fill in the blanks with the words or expressions given below, and change the form where necessary.

One of the most effective applications of smart building is _____ optimize the use of conditioned air. A multi-zone variable air volume (VAV) system could use smart controls to more _____ condition each of the six zones. _____ sensors installed in each office area, VAV box can be _____ to cycle back or shut _____ completely when the corresponding space is _____. If most of the employees are _____ of the office, the smart controls can reduce conditioning to save energy. An example of a less efficient _____ is a whole floor with a constant air volume (CAV) system served by one air-handling unit. _____ this case, control options are limited to cycling back or turning _____ the airflow on the entire floor, and so the flexibility of zone control is lacking.

| to | with | off | out | down |
| alternate | differ | effective | vacant | in |

2. Translate English into Chinese.

(1) It takes an enormous amount of energy to condition air.

(2) These advanced controls can limit HVAC consumption in unoccupied building zones, detect and diagnose faults, and reduce HVAC usage during times of peak energy demand.

(3) The study found approximately 50% electricity savings for RTUs with a three-year payback, even in regions with low electricity prices.

(4) Hotel rooms, which could be considered individual zones, offer a unique application for zone control energy savings.

Part V
Writing Like a Professional

Unit 11   Technical Writing

As a engineer or scientist student, you will be required to write technical reports as part of your degree as well as throughout your career. Examples of such reports include annual environmental reports to regulators, annual reports to shareholders, project proposals, tender documents and journal articles. This part is prepared by Universiti Putra Malaysia for its freshmen in Faculty of Engineering and Science.

The Components of a Report

Depending upon its length and purpose, a technical report may include the following components:

- Title Page.
- Disclaimer.
- Abstract.
- Acknowledgements.
- Contents page.
- List of figures and tables.
- List of symbols and definitions.
- Introduction.
- Main sections and subsections.
- Conclusions.
- Recommendations.
- References.
- Appendices.

Sample Report Content Page

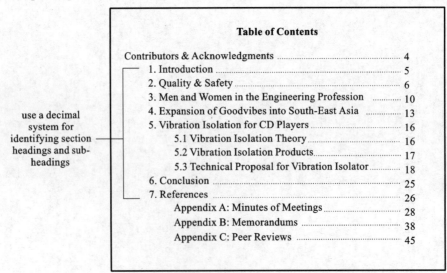

The Layout of a Report

Title Page

The title page will vary according to the style required by the assessor or your company. At a minimum, the title page should include:
- Name of the university.
- Name of school, e. g. School of Mining Engineering.
- Name and code of the subject, e. g. MINE1740 Mining Legislation.
- Title of the report.
- Name of author or authors.
- Date of submission.

Some schools publish styles guides that you are expected to follow when submitting a report. Check with your school office as to whether your school has one.

Figures and Tables

Figures include:
- Diagrams.
- Graphs.
- Sketches.
- Photographs.
- Maps.

Tables represent data in columns.

All figures and tables should be numbered and labelled. Each should have a very simple, descriptive caption explaining the figure or table. Any symbols or abbreviations used in the figure or table must be explained in the text.

The figure must also be referred to in the text, identified by its number, e. g. Fig. XX. Avoid using "the figure above" or "the figure below", as text locations may change when your report is edited. All figures and tables must be referenced if copied or adapted from another source.

Equations and Formulae

Equations should be numbered as they appear in the text, with a number in brackets on the right hand side margin. This number is used for identification throughout the rest of the text.

Equations are generally centred, with consecutive equations on separate lines and with the equal sign ($=$) vertically aligned.

$$y = mx + b \qquad (1)$$
$$x = l(h + f) \qquad (2)$$

Chapter Numbering System

The numbering of chapters and subheadings is normally undertaken throughout the report. The Introduction is generally numbered 1 with the Reference section having the last number. Third level headings are the generally accepted limit (e. g. 8.4.3); too many levels make readers confused.

The preliminary sections (i. e. Table of Contents) prior to the Introduction are not numbered. Appendices are usually labelled with letters, e. g. Appendix B.

Font

Fonts that are easy to read are generally chosen for a report. Times New Roman, Arial and Helvetica are the most popular.

Font size should be a minimum of 12 point for the body text, larger sizes are used for the headings with first level headings being the largest.

The same font should be used throughout the report. It is important not to distract the reader from the contents of the report. Most word processing programs have report templates in them which can be used as a basis for your report style.

Appendices

Appendices are supplements to a report. They are included as separate sections, usually labelled Appendix A, Appendix B etc., at the back of the report. An appendix includes:

1. Information that is incidental to the report.
2. Raw data and evidence which supports the report.
3. Technical data which is too long and or detailed but which supports the report.
4. Maps, folded diagrams, tables of results, letters are some examples.

Technical Writing Conventions

Good technical writing aims to inform with clarity and precision. There are a number of conventions for technical writing and some of the most common are outlined below.

Language and Style

- Aim to inform.

Scientific or technical writing is different from literary writing in a number of ways. Primarily, the aim of technical writing is to inform rather than to entertain. Hence, the style of writing adopted is generally simple and concise.

An example of a literary sentence:	The wind was blowing fiercely and the air outside was growing chilled.
An example of a scientific sentence:	Onshore winds travelling at 45 km per hour brought temperatures down to 15 degrees Celsius.

As informing an audience is the primary aim of the scientific writer, emotive language is avoided. The scientific writer should try to transmit information as objectively as possible.

- Be concise.

Avoid too many long sentences. Sentences with four or more clauses, or parts, are confusing to read. Your text will probably read better if you consider making two sentences rather than one long sentence. If you want to include a qualification or an example then a long sentence is usually appropriate.

An example of a long sentence: After consulting three manufacturers: Dribble and Co., Sooky Ltd. and Bungle Pty, we have found that there are two types of vibration suppression devices for portable CD players and both are simple in design but have inherent drawbacks.

More concise sentence: Three manufacturers were consulted: Dribble and Co., Sooky Ltd. and Bungle Pty. We found two types of vibration suppression devices for portable CD players. Both are simple in design but have inherent drawbacks.

Use words and expressions economically. If you can use one word instead of two or three, then chose the one word.

- Be clear.

Avoid being unclear and ambiguous. This can happen when you do not specify what you are writing about and can even depend how you use words like "it" "this" "thing" "way" "some", etc.

An example of unclear expression: The way we did the experiment was not so successful. Some of what we needed wasn't there.

An example of clear expression: We were unable to complete the experiment. The glass tubing and tripods required for the experiment were not located in laboratory GO25.

Do not use contractions of verbs and pronouns as these are "spoken forms" (doesn't, can't, it's, they're). The formal writing you will do at university and in the workplace will require the full form (does not, can not, it is, they are).

- Be correct.

Check the spelling, punctuation and grammar of your sentences and make sure they are correct. If you use a computer spell checker, be careful. Make sure that you know which word to select. Many easily corrected errors in your written work will affect your presentation and your marks. Sometimes you can see errors more easily if you do not proofread your writing until a day or two after finishing writing. This is called "the drawer treatment". The Learning Centre has many resources on punctuation, grammar and spelling that you can use.

Jargon

Jargon is the technical terminology of any specialised field. Jargon is commonly used when you communicate with others in your field. Communication problems can begin when jargon is used in communications aimed at a more general audience.

Jargon also includes sub-technical words. These have multiple meanings in general and technical contexts. For example the word "fast" has very different meanings in medicine (resistant to), mining (a hard stratum under poorly constructed ground) and painting (colours not affected by light, heat, or damp). A specialist dictionary is required for learning technical and sub technical vocabulary. Your lecturer can recommend a good specialist dictionary.

Aim to write for your intended audience. If your report is for your supervisor or a colleague, then the use of jargon may be both appropriate and expected. If, however, you are writing a report for a general audience or an expert from another field, jargon should be avoided and simple, clear descriptions should be used instead.

Abbreviations and Acronyms

In scientific and technical writing abbreviations and acronyms are commonly used. Abbreviations are pronounced as letters, e.g. UNSW, whereas acronyms are pronounced as words, e.g. LASER. The first time you use an abbreviation or acronym, you must spell out the full term followed by the abbreviation or acronym in brackets. Subsequent use of the term is then made by its abbreviation or acronym.

For example, The University of New South Wales (UNSW) is situated on Anzac Parade, Kensington. The best way to travel to UNSW is by public transport.

The use of an abbreviation is largely dictated by the number of times you are going to be using the term. If the term is only to be used three or four times, it may be better to use the full term each time. This will improve readability, especially if you are using a number of different abbreviations throughout your report.

Using "I" in Technical and Scientific Writing?

There is no single easy answer to this question—it depends. First we recommend that you check with your lecturer/ tutor if and when you can use "I" in your writing.

Reasons for using "I" include:

• The more practised a writer is, the more latitude the writer can have in being casual or creative.

• If a writer is an accomplished engineer/scientist/professional, then as an "expert" in their field the writer can use "I" to give authority to their ideas.

Reasons for not using "I" include:

• When "I" is used too often, it can make your writing sound casual or spoken in style rather than formal and objective.

• Not using "I" can make your writing more believable. The reader may interpret your use of "I" to mean that you are not aware of formal writing conventions. By following conventions you show you are aware of the practices in your field. The reader may also interpret your use of "I" to mean that you are not aware or clear about what other experts in the field have done or think, so instead you are making your own choice.

• In a student's writing using "I" can suggest absorption with the self or that the student does not recognise that their work needs to stand up to scrutiny.

Where Possible Use Active Voice

What is different about these two sentences?

• Male guppies advertise their attractiveness by displaying their colourful patterns (Active Voice).

• Attractiveness is advertised by male guppies by displaying their colourful patterns (Passive voice).

Using active voice in your writing creates a direct and concise message, which also makes your writing easier to read. While we encourage you to use the active voice, this does not mean that you cannot use passive voice, as it can be convenient and necessary. Most writing will have a mixture of active and passive clauses depending on what word is chosen for the subject of a sentence. Look at course related texts that you consider well written to notice how and when writers use active and passive voice.

Non-Discriminatory Language

The use of non-discriminatory language is a legal obligation for all writers. It aims for truthful reporting of the facts. You should avoid statements that suggest bias or prejudice towards any group. You should also avoid making unsupported statements about a person's age, economic class, national origin, political or religious beliefs, race or sex. For example, referring to all persons in an industry as "he" can be inaccurate and misleading. It is best to name the profession using a non-sexist term (e.g.: police officer). (The Learning Centre has a handout "A guide to non-discriminatory language".)

Unit 12   Design Specification

In this age of BIM and information management, it is easy to forget that a key part of information used during construction projects is the specification. It is the mechanism for the designers to impart their technical detailed requirements for the engineering services and systems to the installers and in turn the installer will use it as their scope of work. Specifications which are inaccurate or inconsistent with other project documentation can lead to confusion, errors in pricing and can exacerbate difficulties in the commercial and contractual process.

It is vital, therefore, that specifications are written in such a way as to be readily understandable by the user. They should clearly but unambiguously describe not only what systems are to be provided and how they are intended to operate once installed.

What Is Its Purpose?

The purpose of the specification, as well as the functions described in the foregoing section, is to define the scope of work which is used to form the contractor's duties under the terms of the contract.

How Is It Used?

The specification may be used in many ways throughout the life of a project:

• Prior to tendering, the specification, or parts of it, can be used to demonstrate to the client that their requirements have been met by the design team and accurately passed on to the construction team.

• The specification is used to describe the works to the contractor (initially the tenderer), and to define the legal requirements under the contract they have entered into for the execution of the project.

• The specification can be used as a reference source in times of conflict, to help arbitrate on a particular point, or to provide clarity over an issue. In this respect the specification will be the primary source of legal authority though it may refer to other documents.

• The specification may be read in detail at the start of the project, and then referred to as necessary throughout the project to answer issues as they arise.

• Different sections may be distributed to various sub-contractors or trades as a description of their particular work element where they don't need information on the project as a whole.

• After completion of the project, the specification may be retained as a source of reference for future use by the building occupier, owner or operator, either to compare the performance of the systems with the original design intent, or for information when remodelling or refurbishment is being considered.

What Are the Component Parts?

Information generally common to all specification, regardless of format or arrangement, includes:

• Preliminaries—this section will normally cover general contract conditions and legal issues, but not technical matters.

• Materials and workmanship — this section covers the way plant and systems should be installed, and the quality of plant and materials to be used. This is the quality control aspect of the project and may not be specific to any particular project, but rather more generic in nature.

• Project specific requirements—this is where the particular requirements of the project under consideration are dealt with. Here such information as the project, general description design criteria, descriptions of the systems how they are intended to work, and controls strategies, is all contained. This is unique to each particular project, and quite often is the first part that the tenderer/ constructor will read in order to gain a quick understanding of the project.

• Project specific materials and equipment— this is where detailed information on the particular plant and equipment to be used for the project are located. This information is quite often in the form of schedules, either as part of the specification or as an appendix. The information contained in here will typically deal in quantities, performance data, manufacturer, etc.

The above represents the basic technical elements of a specification. However, it may be beneficial to include other content which has not traditionally been included. The addition of "deliverables" section—information which has to be supplied as part of the tender return—is worthwhile.

When Is It Assembled?

Traditionally, the specification is the last part of the design work to be carried out as much of it cannot be accurately prepared until all aspects of the design have been settled. Some parts, however, can be commenced ahead of the completion of the final design and, where this is possible, the time saved at the end may prove very valuable.

It may be possible during the design phase to prepare some of the system descriptions and design criteria content from other sources such as scheme design reports, thus avoiding duplicated effort. Scheme design report text includes the reasoning behind the selection of systems, something not generally included in specifications. The use of this text can be very helpful as it can give a better understanding of the systems.

How Is It Arranged?

There are a number of ways to arrange a specification, but they generally fall into two approaches:

1. By parts.

2. By work section or system.

In the "by parts" arrangement, the specification is generally arranged by its constituent parts. So, a specification would typically have three or four parts:

Part A: Preliminaries.

Part B: Project specific requirements.

Part C: Project specific materials and equipment.

Part D: Materials and workmanship.

The General Specification For Air-Conditioning, Refrigeration, Ventilation and Central Monitoring & Control System Installation In government Buildings of the Hong Kong Special Administrative Region is an example of this.

Specification Example 1

The "by work section or system" arrangement is often based around the work sections. This format has been used for some of the proprietary specification writing software products on the market, and has all the information about a system contained within a single section. Therefore, the specification would be made up of a number of sections (one for each system, typically) each containing project specific requirements, information on materials and workmanship and project specific materials. This concept is shown in the specification template from a UK design company.

Preparing Effective Specifications

In order to prepare a comprehensive and effective specification, a number of basic principles should be followed. Mostly these are common sense but basic errors can cause difficulties for the recipient in understanding the requirements of specification.

Specification Example 2

The watchpoints for writing specification included in Table 12-1 capture some basic principles, and if they are followed it should result in a clear and effective specification.

Table 12-1　Watchpoints for Writing Specification

Be specific	State clearly what you require the reader to do, or what their responsibilities are. Do not use generalities or vague descriptions
Use the correct level of detail	State clearly what you require the reader to do, or what their responsibilities are. Do no use generalities or vague descriptions
Ensure compliance	Once you have written the specification or the section you have been working on, read it through. Then think from the users' perspective, is it clear what you want the readers to do? Can the readers understand what it is they are required to do? Is the extent of the systems clearly defined, including the operational and control methodologies?
Use the imperative tense	Rather than say "The system shall be installed…" be clearer by using the imperative tense and saying "Install the system…" this takes away doubts or ambiguity over who is doing the installing
Avoid unnecessary duplication	Avoid unnecessary duplication, both between different sections and between sections and schedules etc. Also, there is no need to write a detailed description of something that is shown on a drawing. This reduces the possibility of contradiction or conflict when something changes later
Avoid unnecessary content	The tenderer/constructor is responsible for the specified work as that is what they are employed to do. There is no need to restate this throughout the specification. This, together with the imperative tense, means there is no need for expressions such as "The contractor shall…" and "The contractor shall be responsible for…"

Appendixes Important Words and Phrases

基础科学中英文对照词汇表

absorptivity	吸收率	control mass	控制质量
adhesive forces	附着力	control surface	控制面
adiabatic	绝热	control volume	控制容积
angular velocity	角速度	convection heat transfer	对流换热
atmosphere	大气	counter-flow	逆流
back difference	向后差分	critical point	临界点
black body	黑体	cycle	循环
body-force	质量力	deformation velocity	变形速度
boundary condition	边界条件	density	密度
boundary layer	边界层	differential pressure	差压,压差
buoyant force	浮力	diffuser	扩压管
Carnot cycle	卡诺循环	dimensional analysis	量纲分析
Carnot's theorem	卡诺定理	dimensionless number	无量纲数
closed system	闭口系	dryness	干度
coefficient	系数	dynamic pressure	动压强
combined cycle	联合循环	dynamic viscosity	动力黏度
compressibility	压缩性	efficiency	效率
compressibility factor	压缩因子	elevation head	位置水头
compressible fluid	可压缩流体	emissivity	发射率
condition of stability	稳定性条件	energy	能量
conservation equation of energy	能量守恒方程	engineering thermodynamics	工程热力学
conservation equation of mass	质量守恒方程	enhancement of heat transfer	强化传热
conservation of energy	能量守恒	enthalpy	焓
conservation of mass	质量守恒	entropy	熵
conservation of momentum	动量守恒	equilibrium	平衡
continuity	连续性	external flow	外部流动
continuity equation	连续性方程	field	场

续表

fin	肋片	hydraulic diameter	水力直径
flow boundary layer	流动边界层	hydraulically smooth zero of turbulent pipe	紊流光滑管区
flow pattern	流型		
fluid	流体	ideal gas	理想气体
fluid dynamics	流体动力学	incompressible fluid	不可压缩流体
fluid field	流场	initial condition	初始条件
fluid machinery	流体机械	insulating material	隔热材料
fluid mechanics	流体力学	intensity of turbulence	紊流(强)度
fluid particle	流体质点	intensive quantity	强度量
fluid statics	流体静力学	interface	分界面
forced convection	强制对流	internal combustion engine	内燃机
Fourier number	傅里叶数	internal energy	内能
Fourier's Law	傅里叶定律	internal flow	内部流动
free surface	自由表面	internal friction	内摩擦
friction	摩擦	inversion temperature	转变温度
friction coefficient	摩擦系数	in-viscid fluid	无黏性流体
gas constant	气体常数	irreversible cycle	不可逆循环
gas dynamics	气体动力学	irreversible process	不可逆过程
gauge pressure	表压力	isentropic	定熵(等熵)
governing equation	控制方程	isobaric	定压(等压)
grey body	灰体	isolated system	孤立系
head loss	压头损失	isometric process	定压过程
heat (enthalpy) of formation	生成热(焓)	isothermal	定温(等温)
heat conduction	导热	isothermal surface	等温面
heat exchanger	换热器	isotherms	等温线
heat flux	热流密度	isotropic flow	均质流动
heat of combustion	燃烧热	kinematic energy	动能
heat pipe	热管	kinematic similarity	运动相似性
heat pump	热泵	kinematic viscosity	运动黏度
heat sink	冷源	laminar flow	层流
heat source	热源	latent heat	潜热
heat transfer rate	热流量	linear velocity	线速度
humidity	湿度	liquid	液体

续表

log-mean temperature difference	对数平均温差	parameter	参数
mass	质量	parameter of state	状态参数
mass flow rate	质量流量	perfect gas	理想气体
mass transfer process	传质过程	phase	相
minor loss	局部阻力	pipe flow	管流
mixed convection	混合对流	Planck's Law	普朗克定律
mixture of gases	混合气体	plane flow	平面流动
moist air	湿空气	point of transition	过渡点
moisture content	含湿量	potential energy	位能
move velocity	平移速度	power	功率
multidimensional steady state heat conduction	多维稳态导热	power cycle	动力循环
		Prandtl number	普朗特数
multi-phase flow	多相流流动	pressure	压力
natural convection	自然对流	pressure differential	压差
near wall region	近壁区	pressure drag	压差阻力
Newton's Law of Cooling	牛顿冷却定律	pressure field	压强场
Newton's Viscosity Law	牛顿黏性定律	pressure gage	压强计
Newtonian fluid	牛顿流体	pressure gradient	压强梯度
node	节点	pressure head	压强水头
nonequilibrium	非平衡	process	过程
non-uniform	非均匀	psychrometric chart	焓湿图
non-viscous fluid	非黏性流体	radial velocity	径向速度
normal direction	法向	Rankine cycle	朗肯(兰金)循环
normal line	法线	ratio of pressure of cycle	循环增压比
number of heat transfer unit	传热单元数	ratio of specific heat	比热比
Nusselt number	努塞尔数	real gas	实际气体
one dimensional flow	一维流动	reduced parameter	对比参数
open channel flow	明渠流动	refrigerant	制冷剂
open system	开口系	refrigeration cycle	制冷循环
orifice plate	孔板	refrigerator	制冷机
Otto cycle	奥托循环	regenerative cycle	回热循环
output	输出	reheated cycle	再热循环
parallel flow	层流流动	relative humidity	相对湿度

续表

reversed Carnot cycle	逆卡诺循环	the Second Law of Thermodynamics	热力学第二定律
reversed cycle	逆循环		
reversible process	可逆过程	the Third Law of Thermodynamics	热力学第三定律
Reynolds number	雷诺数		
rotation velocity	旋转速度	thermal boundary layer	热边界层
roughness	粗糙度	thermal coefficient	热系数
saturated	饱和的	thermal conductivity	导热系数
secondary flow	二次流	thermal efficiency	热效率
separation point	分离点	thermal radiation	热辐射
shear deformation	剪切变形	thermal resistance	热阻
shear stress	剪切力	thermal resistance of fouling	污垢热阻
similarity	相似性	thermodynamics	热力学
skin friction	表面摩擦	total pressure	总压强
specific force of gravity	比重	turbulent flow	紊流
Specific Heat	比热	uniform flow	均匀流动
specific humidity	绝对湿度	unit vector	单位矢量
specific volume	比体积	unsteady flow	非定常流
speed of sound	声速	unsteady heat conduction	非稳态导热
standard atmosphere	标准大气压	vacuum degree	真空度
standard state	标准状况	velocity	速度
state	状态	velocity boundary layer	速度边界层
static pressure	静压强	velocity gradient	速度梯度
steady flow	稳定流动	velocity head	速度水头
steam	水蒸气	Venturi flowmeter	文丘里流量计
streamline	流线	vertical force	垂直力
stress	应力	viscosity	黏度
superheated steam	过热蒸汽	viscosity factor	黏度系数
technical work	技术功	viscous fluid	黏性流体
temperature	温度	wet saturated steam	湿饱和蒸汽
temperature field	温度场	wet-bulb temperature	湿球温度
temperature gradient	温度梯度	work	功
the First Law of Thermodynamics	热力学第一定律	working substance	工质

HVACR 中英文对照词汇表

暖通空调设计

英文	中文	英文	中文
absorptance for solar radiation	太阳辐射热吸收系数	heat gain from occupant	人体散热量
additional heat loss	附加耗热量	heat loss by infiltration	冷风渗透耗热量
air humidity	空气湿度	heat storage capacity	蓄热特性
air velocity at work space	工作区空气流速	heating load	热负荷
		indoor air design conditions	室内空气计算参数
atmospheric transparency	大气透明度	indoor air velocity	室内空气流速
barometric pressure	大气压力	indoor temperature(humidity)	室内温(湿)度
building envelope	围护结构	latent heat	潜热
coefficient of heat transfer	传热系数	load pattern	负荷模式
conditioned zone	空气调节区	local solar time	地方太阳时
cooling load	冷负荷	maximum sum of hourly cooling load	逐时冷负荷综合最大值
cooling load from heat conduction through envelope	传热冷负荷		
cooling load from outdoor air	新风冷负荷	mean annual temperature (humidity)	年平均温(湿)度
cooling load temperature	冷负荷温度	mean daily temperature (humidity)	日平均温(湿)度
correction factor for orientation	朝向修正率		
damping factor	衰减倍数	mean monthly maximum/minimum temperature	月平均最高/低温度
days of heating period	采暖期天数		
degree-days of heating period	采暖期度日数	mean monthly temperature (humidity)	月平均温(湿)度
dew-point temperature	露点温度		
direct solar radiation	太阳直接辐射		
dominant wind direction	盛行风向	mean wind speed	平均风速
dry-bulb temperature	干球温度	moisture gain	散湿量
extreme maximum temperature	极端最高温度	outdoor air design conditions	室外空气计算参数
extreme minimum temperature	极端最低温度		
global radiation	总辐射	outdoor critical air temperature for heating	采暖室外临界温度
heat (thermal) lag	延迟时间		
heat flow rate	热流量	outdoor design dry-bulb temperature for summer air conditioning	夏季空气调节室外计算干球温度
heat gain from appliance and equipment	设备散热量		
heat gain from lighting	照明散热量		

续表

outdoor design hourly temperature for summer air conditioning	夏季空气调节室外计算逐时温度	radiation intensity	辐射强度
		relative humidity	相对湿度
		sensible heat	显热
outdoor design mean daily temperature for summer air conditioning	夏季空气调节室外计算日平均温度	shading coefficient	遮阳系数
		sky radiation	天空散射辐射
		solar altitude	太阳高度角
outdoor design relative humidity for summer ventilation	夏季通风室外计算相对湿度	solar azimuth	太阳方位角
		solar constant	太阳常数
outdoor design relative humidity for winter air conditioning	冬季空气调节室外计算相对湿度	solar declination	太阳赤纬
		solar irradiance	太阳辐射照度
		solar radiant heat	太阳辐射热
outdoor design temperature for calculated envelope	围护结构室外计算温度	solar radiation	太阳辐射
		sol-air temperature	综合温度
outdoor design temperature for heating	采暖室外计算温度	space heat gain	房间得热量
		space moisture load	房间湿负荷
outdoor design temperature for summer ventilation	夏季通风室外计算温度	steady-state heat transfer	稳态传热
		temperature at work area	作业地带温度
outdoor design temperature for winter air conditioning	冬季空气调节室外计算温度	temperature difference correction factor of envelope	围护结构温差修正系数
outdoor design temperature for winter ventilation	冬季通风室外计算温度	temperature gradient	温度梯度
		thermal conductivity	导热系数
outdoor design wet-bulb temperature for summer air conditioning	夏季空气调节室外计算湿球温度	thermal diffusivity	导温系数
		thermal resistance	热阻
		unsteady-state heat transfer	非稳态传热
outdoor mean air temperature during heating period	采暖期室外平均温度	ventilation heat loss	通风耗热量
		wet-bulb temperature	湿球温度
percentage of possible sunshine	日照率	wind direction	风向

供暖

angle valve	角阀	building heating entry	热力入口
automatic vent	自动放气阀	by-pass pipe	旁通管
boiler	锅炉	ceiling panel heating	顶棚辐射采暖
boiler room	锅炉房	central heating	集中采暖
branch pipe	支管	circulating pipe	循环管

续表

circulating pump	循环泵	heating equipment	采暖设备
closed return	闭式回水	heating pipe line	采暖管道
condensate pipe	凝结水管	hot water pipe	热水管
condensate tank	凝结水箱	indirect heat exchanger	表面式换热器
convector	对流式散热器	infrared heating	红外线辐射采暖
coupling	管接头		
cross	四通	local heating	局部供暖
direct return system	异程式系统	make-up water pump	补给水泵
district heating	区域供暖	movable support	活动支架
drain pipe	泄水管	natural heating	自然采暖
dry return pipe	干式凝结水管	one-and-two pipe combined system	单双管混合式采暖系统
elbow	弯头		
electric radiant heating	电热辐射采暖	one-pipe cross-over system	单管跨越式系统
expansion pipe	膨胀管		
expansion tank	膨胀水箱	one-pipe series-loop system	单管顺序式系统
feeding branch of radiator	散热器供热支管		
		open return	开式回水
fixed support	固定支架	open tank	开式水箱
float valve	浮球阀	overflow pipe	溢流管
floor heating	地板辐射采暖	pipe fittings	管道配件
gate valve	闸阀	pressure of steam supply	供汽压力
general heating	全面采暖	pressure reducing valve	减压阀
geotherm	地热	radiant heating	辐射采暖
heat capacity	制热量	radiator	散热器
heat emitter	散热设备	radiator valve	散热器调节阀
heat exchange	换热	return branch	回水支管
heat exchanger	换热器	return water temperature	回水温度
heat exchanger	换热器	reversed return system	同程式系统
heat media/medium	热媒	riser	立管
heat source	热源	safety valve	安全阀
		solar heating	太阳能采暖
heat transportation & distribution network	热网	steam boiler	蒸汽锅炉
		steam ejector	蒸汽喷射器

续表

steam pipe	蒸汽管	thermostat	恒温计
steam trap	疏水器	two-pipe system	双管采暖系统
steam-water heat exchanger	汽-水式换热器	vent pipe	排气管
		vertical one-pipe system	垂直单管系统
stop valve	截止阀	water-water heat exchanger	水-水式换热器
supply water temperature	供水温度		
thermometer	温度计	wet return pipe	湿式凝结水管

通风

access door	检查门	fan	风机
air balance	风量平衡	fire damper	防火阀
air changes	换气次数	fittings	（通风）配件
air intake	进风口	flexible duct	软管
axial fan	轴流式通风机	flexible joint	柔性接头
back-flow preventer	防回流装置	ganopy hood	伞形罩
butterfly damper	蝶阀	general ventilation	全面通风
by-pass damper	旁通阀	guide vane	导流板
capture velocity	控制风速	heat balance	热平衡
ceiling fan	吊扇	heat release	散热量
centrifugal fan	离心式通风机	heat screen	隔热屏
circulating fan	风扇	hood	局部排风罩
cleanout opening	清扫孔	inductive ventilation	诱导通风
combined ventilation	联合通风	local ventilation	局部通风
cross-flow fan	贯流式通风机	louver	百叶窗
cross-ventilation	穿堂风	mechanical ventilation	机械通风
diffuser	散流器	natural ventilation	自然通风
emergency ventilation	事故通风	opening	开口
exhaust air rate	排风量	pressure relief device	泄压装置
exhaust fan	排风机	self-contained cooling unit	自备冷风机组
exhaust fan room	排风机室	smoke control	防烟
exhaust opening	吸风口	smoke damper	防烟阀
exhaust system	排风系统	smoke exhaust damper	排烟阀
exit	排风口	smoke extraction	排烟
face velocity	罩口风速	source of heat release	散热源

续表

supply air rate	进风量	ventiduct (duct)	通风管
supply fan	送风机	ventilate	使通风
supply system	送风系统	ventilated	通风的
unidirectional flow ventilation	单向流通风	ventilated roof	通风屋顶
unorganized air supply	无组织进风	ventilation rate	通风量
unorganized exhaust	无组织排风	ventilator	通风机
unventilated	不通风的	zone of negative pressure	负压区
vent	通风孔	zone of positive pressure	正压区

空气调节

absorption refrigeration cycle	吸收式制冷循环	constant volume system	定风量系统
		cooler	冷藏器
accessories	附件	cooling & heating sources	冷热源
adiabatic humidification	绝热加湿	cooling equipment	制冷设备
air conditioning system	空调系统	cooling tower	冷却塔
air distribution	气流组织	cooling water	冷却水
air preheater	空气预热器	degree of subcooling/heating	过冷/热度
air-cooled condenser	风冷式冷凝器	dehumidification	减湿
air-water system	空气-水系统	dehumidifying cooling	减湿冷却
all-air system	全空气系统	diffuser	散流器
all-water system	全水系统	discharge temperature	排气温度
angle scale	热湿比	dry cooling condition	干工况
Carnot cycle	卡诺循环	dual duct system	双风管系统
central system	中央空调	duct system	风管系统
centrifugal compressor	离心式压缩机	ductwork	风管网路
chilled water	冷冻水	electric heater	电加热器
chiller	制冷机	evaporating pressure	蒸发压力
coefficient of performance	性能系数	evaporating temperature	蒸发温度
compression-type refrigerating machine	压缩式制冷机	evaporator	蒸发器
		fan section	风机段
condensate drain pan	凝结水盘	filter section	过滤段
condensation	冷凝	forward flow zone	射流区
condenser	冷凝器	free jet	自由射流
condensing temperature	冷凝温度	fresh air handling unit	新风机组

续表

fresh air requirement	新风量	refrigerator	冰箱
heat and moisture transfer	热湿交换	return air	回风
heat pipe	热管	return fan	回风机
heating equipment	制热设备	return flow zone	回流区
heating/cooling coil	热/冷盘管	reverse Carnot cycle	逆卡诺循环
humidifier	加湿器	screw compressor	螺杆式压缩机
humidity	湿度	secondary return air	二次回风
humidity ratio	含湿量	sensible cooling	等湿冷却
isothermal humidification	等温加湿	sensible heating	等湿加热
isothermal jet	等温射流	shell and tube condenser	壳管式冷凝器
mixing box section	混合段	sidewall air supply	侧面送风
modular air handling unit	组合式机组	single duct system	单风管系统
muffler section	消声段	specific enthalpy	比焓
nozzle outlet air supply	喷口送风	split unit	分体式空调
outlet air velocity	出口风速	spread	射流扩散角
percentage of return air	回风百分比	subcooling	过冷
perforated ceiling air supply	孔板送风	suction pressure	吸气压力
pipework	管道网路	suction temperature	吸气温度
piping system	管道系统	supply air	送风
plate heat exchanger	板式换热器	terminal unit	末端设备
primary return air	一次回风	thermostatic expansion valve	热力膨胀阀
psychrometric chart	焓湿图	three-pipe water system	三管制水系统
rating under air conditioning condition	空调工况制冷量	throttling expansion	节流膨胀
		throw	射程
reciprocating compressor	活塞式压缩机	two-pipe water system	两管制水系统
refrigerant	制冷剂	variable air volume system	变风量系统
refrigerate	制冷	wall attachment jet	贴附射流
refrigerating compressor	制冷压缩机	water-cooled condenser	水冷式冷凝器
refrigerating station	制冷机房	wet cooling condition	湿工况
refrigeration system	制冷系统	window unit	窗式空调

给排水中英文对照词汇表

absolute roughness	绝对粗糙度	effective coefficient of local resistance	折算局部阻力系数
aluminium pipe	铝管		
asbestos cement pipe	石棉水泥管	effective length	折算长度
available pressure	资用压力	elbow	弯头
average daily coefficient	平均日供水量	equivalent coefficient of local resistance	当量局部阻力系数
branch pipe	支管		
branch system	枝状管网	equivalent length	当量长度
bushing	补心	expansion pipe	膨胀管
buttress	水管支墩	feeding branch	供热支管
by-pass pipe	旁通管	fire demand	消防用水
cast-iron pipe	铸铁管	fixed support	固定支架
cement pipe	混凝土管	fresh water	淡水
channel	渠道	friction factor	摩擦系数
circuit; loop	环路	friction loss	摩阻损失
circulating pipe	循环管	glass pipe	玻璃管
clay pipe	陶土管	green belt sprinkling	绿化用水
close nipple	螺纹接口	grey cast-iron pipe	灰口铸铁管
coating pipe	涂敷管	ground water	地下水
composite pipe	复合管	head loss	水头损失
condensate pipe	凝结水管	heating pipe line	采暖管道
cooling water	冷却水	hot water pipe	热水管
copper pipe	铜管	hourly variation coefficient	时变化系数
corrugated pipe	波纹管	hydraulic calculation	水力计算
coupling	管接头	hydraulic disorder	水力失调
cross	四通	hydraulic resistance balance	阻力平衡
daily variation coefficient	日变化系数	index circuit	最不利环路
distribution system	配水管网	lead pipe	铅管
domestic water	生活用水	leakage	管网漏失水量
drain pipe	泄水管	length of pipe section	管段长度
drainage pipe	排污管	limiting velocity	极限流速
dry return pipe	干式凝结水管	lining pipe	衬敷管
economic velocity	经济流速	local resistance	局部阻力

续表

main pipe; trunk pipe	干管	sludge disposal	污泥处置
maximum service coefficient	最高日供水量	sludge treatment	污泥处理
metal flexible hose	金属软管	specific frictional resistance	比摩阻
minimum service head	最小服务水头	static pressure	静压
movable support	活动支架	steam pipe	蒸汽管
nodular cast-iron pipe	球墨铸铁管	steel pipe	钢管(铁管)
non-common section	非共同段	storage reservoir	贮水池
once through system	直流水系统	street flushing demand	浇洒道路用水
operating range	作用半径	surface water	地表水
output	供水量	system resistance	系统阻力
overflow pipe	溢流管	tee	三通
pipe fittings	管道配件	thermal insulation pipe	保温管
pipe network	管网	total pressure	全压
pipe section	管段	unforeseen demand	未预见用水量
plastic pipe	塑料管	union	活接头
pressure drop	压力损失	velocity pressure	动压
process water	生产用水	vent; vent pipe	排气管
pumping house	泵站	wastewater	废水
raw water	原水	wastewater disposal	废水处置
reducing coupling	异径管接头	wastewater flow	污水量
reinforced concrete pipe	钢筋混凝土管	wastewater flow norm	排水定额
relative roughness	相对粗糙度	water consumption	用水量
return branch	回水支管	water consumption norm	用水定额
riser	立管	water quality	水质
rubber pipe	胶管	water reuse system	复用水系统
screw nipple	丝对	water reuse system	循环水系统
screwed plug; plug	丝堵	water source	给水水源
sewage treatment	污水处理	water supply system	给水系统
sewage	污水	water treatment	给水处理
sewerage system	排水系统	water-gas steel pipe	水煤气钢管
sludge	污泥	working pressure	工作压力

电气工程中英文对照词汇表

actuator	执行机构	critical voltage	临界电压
adapter	接头	current	电流
adaptive control system	自适应控制系统	current limiter	限流器
		current transformer	电流互感器
air circuit breaker	空气断路器	cut-off current	切断电流
alarm signal	报警信号	D. C contactor	直流接触器
ammeter	电流表	delay	延时
automatic control	自动控制	detecting element	检测元件
automatic power regulator	自动功率调整器	differential pressure type flowmeter	差压流量计
automatic regulator	自动调节器	direct current (D. C)	直流
balanced load	平衡负荷	direct digital control (DDC) system	直接数字控制系统
bayonet socket	卡口插座		
breakdown voltage	击穿电压	distance between lines	线间距离
cable cabinet	电缆箱,分线盒	distribution cabinet	配电箱
		distribution substation	配电站
calculated capacity	计算容量	earth electrode	接地极
camera	摄像头	earthing	接地
capacitance	电容	earthing for lightning protection	防雷接地
centralized control	集中控制	electric discharge	放电
closed switch	封闭式开关	electric shock	触电
communication line	通信线路	electromagnetic starter	电磁起动器
constant value control	定值调节	emergency lighting line	应急照明线
control device	控制装置	emergency source	应急电源
control mechanism	操动机构	fault current	故障电流
control panel	控制屏(台)	feedback	反馈
control valve	调节阀	floating control	无定位调节
control voltage	操作电压	frequency	频率
controlled variable	被控参数(变量)	full-load	满载
		gap	间隙
controller; regulator	调节器	general distribution box	总配电箱
correcting unit	执行器;校正装备	generator	发电机
		ground bus	接地干线

续表

Hertz	赫兹	pressure switch	压力开关
high tension (H.T.)	高压	pressure thermometer	压力式温度计
impedance	阻抗	proportional band	比例带
induction motor	感应电动机	proportional control	比例调节
input variable	输入变量，输入值	protection & earthing	保护和接地
		radial system	放射式
insulating tape	绝缘包布	rated capacity	额定容量
L.T. line	低压线路	rated voltage	额定电压
leakage	漏电	rectifier	整流器
lighting distribution box	照明配电箱	reducer	减压器
lightning protection	防雷	relay	继电器
lightning rod	避雷针	remote control	遥控
lightning surge	雷涌	reset	复位
limiting current	极限电流	resistance	电阻
low tension (L.T.)	低压	selector	选择器
main power line	电力干线	sensor	传感器
main switch	主开关	sequence control	程序控制
monitor	监控	series reactor	串联电抗器
neutral	中性线（N线）	server	服务器
neutral line	零线	set point	给定值
nominal current	额定电流	socket outlet	插座
nominal load	额定负载	solenoid valve	电磁阀
open loop control	开环控制	stabilizer	稳压器
open switch	开启式开关	starter	起动器
open wire	明线	starting controller	起动控制箱
output variable	输出变量	step-up transformer	升压器
overload	过载	supply voltage	供电电压
parameter detection	参数检测	switch	开关
phase	相，相位	switch on	合闸
plug	插头	synchronizer	同步器
plug device	插接装置	transformer	变压器
pole length	电杆长度	uninterrupted power supply (UPS)	不停电电源
positioner	定位器		
power	功率	universal meter	万用电表
power load	动力负荷	voltmeter	电压表

References

[1] Cengel Y A, Turner R H, Smith R. Fundamentals of Thermal-fluid Sciences [M]. New York: McGraw-Hill, 2001.

[2] Cengel Y A, Ghajar A J. Heat and Mass Transfer: Fundamentals & Applications [M]. 4th ed. New York: McGraw-Hill, 2011.

[3] Chadderton D V. Building Services Engineering [M]. 6th ed. London: Routledge, 2012.

[4] Silva V, Hall M, Azevedo I. Low Carbon Transition — Technical, Economic and Policy Assessment[M]. London: IntechOpen, 2018.

[5] Tzimas E, Moss R L, Ntagia P. Technology Map of the European Strategic Energy Technology Plan (SET-Plan) [M]. Luxembourg: Publications Office of the European Union, 2011.

[6] Emmerich S J, Mcdowell T. Initial Evaluation of Displacement Ventilation and Dedicated Outdoor Air Systems for U. S. Commercial Buildings [S]. Washington D C: NISTIR, 2005: 3-7.

[7] Mohsen S, Kandelousi S. HVAC System[M]. London: IntechOpen, 2018.

[8] Sahlot M, Riffat S. Desiccant Cooling Systems: A Review[J]. International Journal of Low-Carbon Technologies, 2016(11): 489-505.

[9] 蓝梅. 给排水科学与工程专业英语[M]. 北京: 化学工业出版社, 2013.

[10] 刘剑. 电气工程及其自动化专业英语[M]. 2版. 北京: 中国电力出版社, 2004.

[11] Greeno R, Hall F. Building Services Handbook [M]. Oxford: Elsevier, 2013.

[12] Hussain Z, Memon S, Shah R, et al. Methods and Techniques of Electricity Thieving in Pakistan[J]. Journal of Power and Energy Engineering, 2016 (4): 1-10.

[13] Blume S. Electric Power System Basics for the Nonelectrical Professional [M]. New Jersey: John Wiley & Sons, Inc, 2017.

[14] Sands J. Model Format for Building Services Specifications(2nd Edition)[S]. London: BSRIA, 2016: 27-33.

[15] Universiti Putra Malaysia. Reading-material Technical Writing [A/OL]. [2021-04-04]. https://www.coursehero.com/file/25118373/Reading-material-twopdf/

[16] King J, Perry C. Smart Buildings: Using Smart Technology to Save Energy

in Existing Buildings [A/OL]. [2021-04-04]. https://www.aceee.org/sites/default/files/publications/researchreports/a1701.pdf

[17] Control Solution Inc. The Ultimate Guide to Building Automation [A/OL]. [2021-04-04]. https://controlyourbuilding.com/blog/entry/the-ultimate-guide-to-building-automation

[18] Adams Electric Cooperative. Grounding & Bonding — Why It Is done and How to Install Properly[A/OL]. [2021-04-04]. http://www.adamselectric.coop/wp-content/uploads/2015/02/Grounding-Bonding.pdf

[19] E2Energy. Electrical Power Systems in Buildings [A/OL]. [2021-04-04]. http://f2energy.com/electrical-power-systems-in-buildings/

[20] Dickenson M. Condensate Drainage[A/OL]. [2021-04-04]. https://www.edtengineers.com/blog-post/condensate-drainage

[21] Bhatia A. Air conditioning psychrometrics[A/OL]. [2021-04-04]. https://www.cedengineering.com/userfiles/Air%20Conditioning%20Psychrometrics%20R1.pdf

[22] Bhatia A. HVAC-Space Heating Systems[A/OL]. [2021-04-04]. https://www.cedengineering.com/userfiles/HVAC%20Space%20Heating%20Systems%20R1.pdf